M000273099

The Astronaut's Cookbook

Tales, Recipes, and More

By Charles T. Bourland
and Gregory L. Vogt

Springer

Charles T. Bourland
1105 NE. 450 Road
Osceola, MO, 64776
USA
cbourlan@dishmail.net

Gregory L. Vogt
2670 Holly Hall Apt H
Houston TX 77054
USA
gregorylvogt@sbcglobal.net

ISBN 978-1-4419-0623-6   e-ISBN 978-1-4419-0624-3
DOI 10.1007/978-1-4419-0624-3
Springer New York Dordrecht Heidelberg London

Library of Congress Control Number: 2009933620

© Springer Science+Business Media, LLC 2010
All rights reserved. This work may not be translated or copied in whole or in part without the written permission of the publisher (Springer Science+Business Media, LLC, 233 Spring Street, New York, NY 10013, USA), except for brief excerpts in connection with reviews or scholarly analysis. Use in connection with any form of information storage and retrieval, electronic adaptation, computer software, or by similar or dissimilar methodology now known or hereafter developed is forbidden.
The use in this publication of trade names, trademarks, service marks, and similar terms, even if they are not identified as such, is not to be taken as an expression of opinion as to whether or not they are subject to proprietary rights.

Printed on acid-free paper

Springer is part of Springer Science+Business Media (www.springer.com)

This book is dedicated to the astronauts who lost their lives pursuing their dreams to explore the frontier of space.

Apollo 1, January 27, 1967
    Virgil "Gus" Grissom
    Edward H. White II
    Roger B. Chaffee

Shuttle STS-51L Challenger, January 28, 1986
    Francis R. (Dick) Scobee
    Michael J. Smith
    Ellison S. Onizuka
    Judith A. Resnik
    Ronald E. McNair
    Sharon Christa McAuliffe
    Gregory Jarvis

Shuttle STS-107 Columbia, February 1, 2003
    Rick D. Husband
    William C. McCool
    Michael P. Anderson
    David M. Brown
    Kalpana Chawla
    Laurel B. Clark
    IIan Ramon

FIGURE 1 Lost mission patches.

Acknowledgements

The authors acknowledge the assistance of Vickie Kloeris, Kimberly Glaus-Late, and Donna Nabors at the NASA Johnson Space Center Space Food Facility with the space food specifications and food standards from which these recipes were derived. The authors also acknowledge Mike Gentry and Adam Caballero of the NASA Johnson Space Center Media Resource Center for help with the NASA photographs.

Contents

About the Authors

Charles T. Bourland spent 30 years at the NASA Johnson Space Center developing food and food packages for spaceflight. He began his work during the *Apollo 12* program and continued through the early years of the International Space Station. During his career at NASA he was involved in the *Apollo* recovery ship food, Zero G testing aboard the Zero G plane, quarantine food systems, and planetary-based food systems for the Apollo program, Skylab, the Apollo-Soyuz Test Program, the shuttle program, and the International Space Station.

Gregory L. Vogt is a veteran writer, science consultant, and developer of science and technology materials for schools. A former science teacher himself, Vogt has been director of an interactive science museum, an education specialist with the Astronaut Office at the NASA Johnson Space Center, and a consultant to museums, Challenger Learning Centers across the United States, and foreign space agencies. Author of more than 80 trade books, Vogt's latest book is a newly published Springer title called *Landscapes of Mars: A Visual Tour*. Currently, he works at the Center for Educational Outreach at the Baylor College of Medicine in Texas.

The Astronaut's Cookbook

CHAPTER 1

Introduction

Half a century has passed since humans began the conquest of space. Satellites, lunar landings, space stations, robot rovers on Mars, solar and deep space observatories, and probes to the edges of interstellar space have sent back a flood of scientific information. Space exploration has fundamentally changed our lives, from the classroom to the marketplace to cyberspace.

Space scientists and mission planners will tell you that the exploration of space is vital to our economic and environmental survival and essential for our security in a dangerous world. We've heard all this before, but what really convinces us that we should go into space is that exploration is just plain exciting, and much of that excitement is in the details. . .

+ What is it like to fly in space?
+ Why do you float?
+ Can you see any manmade objects from space?
+ How does it feel to travel 25 times the speed of sound?

Of course, there are the more mundane questions. How to go to the bathroom in space is tops on the "enquiring minds want to know" list. But after that, the next most popular questions have to do with space food—how do you eat it, what do you eat, how does food taste in space, how do you cook it, and so on.

C.T. Bourland, G.L. Vogt, *The Astronaut's Cookbook*, DOI 10.1007/978-1-4419-0624-3_1,
© Springer Science+Business Media, LLC 2010

For many of us, especially those of us old enough to have been around at the beginning of the space age, our concept of space food is limited to Tang, food sticks, goop squeezed out of toothpaste tubes, and freeze-dried ice cream. We remember old science fiction movies where intrepid space explorers dined entirely on nutrient-packed pills.

We also remember space experts speculating on the problems the first astronauts might encounter in space. Continuous floating inside a spaceship might cause astronauts to go insane from the stress of the sensation of falling and waiting for the impact that never comes. Astronauts could get fried by cosmic radiation. It might not be possible to swallow food, making long space missions impossible.

Fortunately, none of those concerns proved true. Astronauts delight in the floating effects and don't go crazy in space. While in Earth orbit, they are protected from radiation by the Van Allen radiation belts, an early discovery of the *Explorer* satellite program. Eating and swallowing turned out to be easy. This was proved during the second manned spaceflight, in 1961, when Soviet cosmonaut Gherman Titov became the first human to consume food in space. A few months later, John Glenn, Jr., became the first American to consume food in space. He ate applesauce dispensed from a squeeze tube during his February 1962 *Mercury* flight. The experiences of these early space explorers began what is, to this day, an ongoing process of space food development.

Space food is a unique branch of food and nutrition science. It is far more than just selecting tasty and healthy things to eat. Creating space food is also about packaging, preparation, consumption, and disposal. The primary driving force behind space food development is weight and volume. The less the total payload carried by a rocket weighs, including the weight of the astronauts, the less thrust the rocket has to generate to reach space. In relation to volume, the less space occupied by space supplies and tools needed by the crew, the more room there is in the capsule for the crew.

Especially in the early days of spaceflight, everything placed on board for liftoff was measured to the last fraction of an ounce and to the last cubic inch. Space food was no exception. In Project Gemini, the two astronauts jammed into the *Gemini* capsule were each allowed 1.7 lb of food per day. Because of the tightness, it was a Herculean task to provision the *Gemini 7* mission, where astronauts Frank Borman and Jim Lovell remained in close quarters for 14 days in space.

Today's space shuttle crews are allowed 3.8 lb of food per person. The difference in food weight between the two spacecraft has to do with water content. The *Gemini* spacecraft was a tight fit. Astronauts liked to joke, "You don't climb into a space capsule. You put it on!" John Young, one of two astronauts for the first *Gemini* mission, likened the capsule to "sitting in a phone booth that was lying on its side." Because interior space was at a premium, food for the *Gemini* crew had to be as compact as possible. Most *Gemini* foods were dried for launch and rehydrated in space from the spacecraft's water supply.

The space shuttle, flown for the first time 15 years after the last *Gemini* mission, is a much larger spacecraft than the *Gemini* capsule. The orbiter payload bay alone could hold three *Gemini* capsules end to end. Though carrying up to eight astronauts at a time, the shuttle crew cabin has plenty of room for food for all the crew for a two-week mission. The great lifting power of the shuttle's engine permitted food scientists to leave some of the natural water in space shuttle foods, which made them taste better and easier to prepare (Figure 1.1). This is why daily rations for shuttle crews could weigh more than *Gemini* rations.

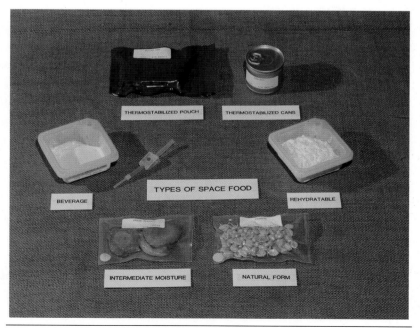

FIGURE 1.1 Early Shuttle foods (NASA photograph).

Space foods are individually packaged and stowed for easy handling in space. All food is precooked or processed, so it requires no refrigeration. It is either provided in a ready-to-eat form, or it can be prepared simply by adding water or by heating.

Because of significant safety issues, astronauts do not really cook food in space. A spacecraft is a closed environment. If an astronaut burns a steak, he or she cannot open the window and let the smoke out. Smoke contains toxins and is a serious health hazard in a closed environment. If the smoke never gets out, it circulates until its soot particles are deposited on every bit of interior surface, including the insides of astronauts' lungs.

Modern space menus also include a small amount of fresh fruits and vegetables that are stowed in a fresh food locker. Without refrigeration, carrots and celery must be eaten within the first two days of the flight, or they will spoil. Other fresh items include tortillas, which have an exceptionally long shelf life. Preferred over sliced bread, tortillas do not easily break down into crumbs that would clog air vent filters and get sucked up into astronauts' nostrils.

Space Food Types

Space food falls into several categories, depending mostly upon how astronauts prepare the food at meal time. In each category, making food easy to handle is of prime importance. Astronaut time in orbit is extremely valuable, and the less time involved in meal preparation and clean up, the more time is available for accomplishing the goals of the mission.

Rehydratable Food

Rehydratable items include both foods and beverages. Water is removed during Earth processing, making it easy to stow the foods and extend the shelf life. During flight, water is added back to the food.

Regardless of how this sounds at first, the strategy of rehydrating food in space actually saves launch weight. How? Why should it make a difference if you send up the food with its water or dry it first and add the water back later?

The space shuttle orbiter, like the *Apollo* and *Gemini* spacecraft previously, generates electricity with fuel cells. Fuel cells combine hydrogen and oxygen to make electricity. Water is a byproduct of the generation process. Since you have to send up the hydrogen and oxygen for power anyway, why not use fuel cell wastewater for rehydrating food and drinking? This means it is unnecessary to launch more than a starter supply of water. In no time, there is plenty of water for food preparation!

Rehydratable foods are packaged in containers that have some sort of port through which water can be added. A label indicates how much water is to be added and whether the water should be hot or cold. The crew member doing the "cooking" inserts the water using a large gauge needle and then kneads the package for a moment to spread the water around so it will make contact with all parts of the food. If need be, the package can be placed in a convection oven to raise its temperature beyond that provided by the hot water. When the food item is ready, it is consumed directly from the package (Figure 1.2).

Rehydratable foods include chicken consommé, cream of mushroom soup, macaroni and cheese, chicken and rice casseroles, shrimp cocktail, and various breakfast foods such as

FIGURE 1.2 Shuttle rehydratable food (NASA photograph).

scrambled eggs and cereals. Breakfast cereals are prepared by packaging the cereal with nonfat dry milk and sugar. Water is added to the package to rehydrate the milk just before the cereal is eaten.

Fewer rehydratable foods are consumed on the International Space Station (ISS) than on the space shuttle. The ISS generates power with huge solar panels that make electricity directly from sunlight. Water is not a byproduct, and consequently, all water is brought to the ISS by the space shuttle or by the Russian *Progress* resupply spacecraft. ISS astronauts must be careful with their water use, and they stretch water supplies by recycling what they can. Thus, there is little advantage in launching large quantities of rehydratable foods.

Thermostabilized Food

Thermostabilized food refers to canned food, like your standard canned peas, beans, or artichokes. The foods are heat-processed to destroy deleterious microorganisms and enzymes. Once made specifically for spaceflight, individual servings of thermostabilized foods are now commercially available in flexible pouches. (Military MRE's, or meals ready to eat, are staples of spaceflight.)

Most of the fish, such as tuna and salmon, and fruit are carried into space in thermostabilized cans or pouches. The cans open with full-panel pull-out lids. Puddings are packaged in plastic cups with pull-off foil lids. Most of the entrees are packaged in flexible retort pouches, similar to MREs. These include products such as grilled chicken, tomatoes, eggplant, beef with barbecue sauce, and ham. After the pouches are heated in the onboard convection oven, the food is cut open and eaten directly with conventional eating utensils. The only space food utensil usually not found on Earth dining tables is a scissors for opening the packages.

Intermediate Moisture Foods

Intermediate moisture foods are those preserved by restricting the amount of water available for microbial growth, while retaining sufficient water to give the food a soft texture. Examples are dried peaches, pears, apricots, and beef. Except for cutting open the package, no preparation is needed. Intermediate moisture foods usually

FIGURE 1.3 Shuttle/ISS dried peaches (NASA photograph).

range from 15 to 30% water, but the water is chemically bound with the sugar or salt (Figure 1.3).

Natural Form Foods

Nuts, granola bars, M&Ms™, and cookies are classified as natural form foods. They are packaged ready to eat in flexible pouches.

Irradiated Meat

Irradiated meat includes beefsteak, fajitas, breakfast sausage, and smoked turkey. To insure long shelf life at the ambient temperatures found inside the spacecraft, the meat is cooked, packaged in flexible foil-laminated retort pouches, and sterilized by zapping it with ionizing radiation (Figure 1.4).

Condiments

Condiments include commercially packaged individual pouches of catsup, mustard, mayonnaise, taco sauce, and hot pepper sauce. Polyethylene dropper bottles contain liquid pepper and liquid salt. The pepper is suspended in oil and the salt is dissolved in water. Drops are pressed directly onto the food.

FIGURE 1.4 Irradiated Smoked Turkey (NASA photograph).

Liquefying pepper and salt may seem strange, but the space-flight environment requires it. Shake out dry pepper and salt on Earth, and it falls onto your food. Doing the same thing in space would create a "cloud" of seasoning, leading to a sneeze-fest and very irritated eyes and nasal passages.

Although salt and pepper can boost the taste of various foods, it is challenging to apply the liquids properly. Shaking salt and pepper on Earth uniformly spreads out the particles. Pressing liquid salt and pepper drops directly on foods in space often leads to "hot spots."

Space Food Menu Development

Developing space food menus is challenging. One of the obvious objectives is to have food that the astronauts will eat. If not eaten, it doesn't matter how healthy and nutritious the food is. Good-tasting food means happy astronauts. Bland or bad-tasting food means leftovers. In space, leftovers are *bad*. Even if a spacecraft has a refrigeration unit, storage space is very limited. Eventually, moldy refrigerator science projects have to be disposed of. The trouble with spaceflight is that you can't pitch the garbage bag out the door. It has to be held until it can be returned to Earth.

Step one in food menu development is to create a potential list of foods. Items on the list have to be available through current food processing technologies. They also have to meet certain constraints imposed by spacecraft, crew members, and the flight environment. For example, a rehydratable food item has to be rehydratable even with cold water in case the hot water supply fails. The food has to have a long shelf life. It must have a good nutritional balance and be available at a reasonable cost. Finally, it has to taste good. That means more than just flavor. It has to have a pleasing texture and color. (The psychology of food is important. Try dyeing a glass of milk green and see how many people will drink it!)

After the food list is settled upon, dietitians extract specific foods from the list to create breakfast, lunch, and dinner menus. They also make recommendations for improvements and add items for balance.

To ensure the food items will be accepted by flight crews, each is tested in a food laboratory. Flight crews are invited to lunch at the lab. With hungry astronauts sitting around a counter, food technicians prepare samples of all the foods to be considered for flight. With pencil and paper, the crew gives every item a score of from 1 to 9. Unlike golf, a low score is bad. A score of 1 means "dislike extremely." A 9 means "like extremely." Crew members not only rate taste but also texture and appearance. Foods must receive an overall score of 6 or better for further consideration.

Dietitians use the ratings to establish a preliminary standard menu. Astronauts are free to pick the standard menu or create their own menu from the approved choices open to them. A crew member may also choose a menu he or she used on a previous space mission.

After menus are selected, crew members try them out during the frequent training simulations. Ground-based trainers that look identical to the real space shuttle orbiter are used for practicing every step and action that will take place during the real mission.

Simulations can take days, and the crew has to eat. It is a great time to test the menus and preparation techniques. Using the same food during simulations is especially beneficial to the food laboratory staff. Having the astronauts involved in choosing their menus significantly reduces the complaints!

The final menus are submitted to a dietitian, who checks them for compliance with established medical requirements. Calories are one of the first items to be evaluated. The number of calories that are required are based on the World Health Organization (WHO) formula:

Men   18–30 years: $1.7(15.3 W*+679)$ = kcal/day required
           30–60 years: $1.7(11.6 W+879)$ = kcal/day required

Women   18–30 years: $1.6(14.7 W+496)$ = kcal/day required
           30–60 years: $1.6(8.7 W+879)$ = kcal/day required

*W = weight in kg.

Calories must be divided along the following guidelines:

Protein = 12–15%
Carbohydrates = 50–55%
Fat = 30–35%

NASA nutritional requirements are the same for both male and female astronauts. This varies from the National Academy of Sciences Dietary Reference Intakes (DRI), which recommends slight variations in diet based upon gender. Vitamin and mineral requirements for astronauts are basically the same as the DRI's, with the exception of iron. Astronauts are limited to 10 mg/d versus the DRI recommendation of 8 mg/d for men and 18 mg/d for women. This is a response to one of the physiological changes that occur in microgravity. Microgravity causes a slight fluid shift in the upper body. Body systems interpret this shift as an excess of blood and increase fluid excretion until a new balance is achieved, resulting in a total blood volume decrease while in flight. Once balance is achieved, red blood cell turnover slows. The need for iron, an oxidant, is reduced.

Another significant physiologic change occurring in microgravity is increased bone loss. Weight-bearing bones lose between 1 and 2% of their calcium mass per month. This amounts to a total loss of approximately 10% of their skeletal calcium in the weight-bearing bones on a 5–6 month mission. The loss is due to the lack of stress placed on the skeleton on a normal day on Earth. The bone loss in 1 year in space is equivalent to the loss a typical person would experience on Earth over a 10-year period starting at age 50. Bone loss could be a "show stopper" for space missions to Mars that could last 2 or more years.

One would think the solution to this problem would be simple—increase calcium by giving astronauts calcium supplements during flight. This seems like it would be a good idea, but it doesn't work. Without skeletal stress, bone-building cells do not capture the extra calcium. Instead, the calcium is excreted through urine, increasing the potential for kidney stones. NASA has studied the problem for many years and found that exercise helps to slow calcium loss. Some drug therapies show promise, but the problem is not yet completely solved.

What does a space menu generally look like? The one below was typical of what the *Apollo* astronauts ate during their Moon missions.

DAY 1

Meal A
Peaches (R)
Bacon Squares (IMB)
Cinnamon Bread Toast Cubes (DB)
Breakfast Drink (R)

Meal B
Corn Chowder (R)
Chicken Sandwiches (DB)
Coconut Cubes (DB)
Sugar Cookie Cubes (DB)
Cocoa (R)

Meal C
Beef and Gravy (R)
Brownies (IMB)
Chocolate Pudding (R)
Pineapple-Grapefruit Drink (R)

ABBREVIATION KEY:
R = Rehydratable
DB = Dry Bite
IMB = Intermediate Moisture Bite

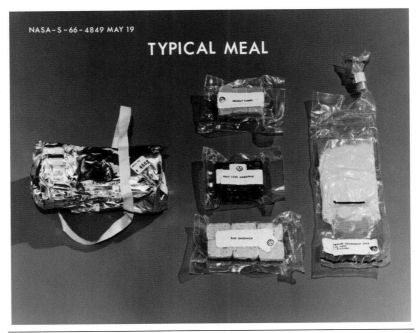

FIGURE 1.5 Gemini Food. Each meal component was over-wrapped in foil. Apollo meals were also over-wrapped in foil (NASA photograph).

All foods were naturally dry or required the addition of water prior to consumption. This made meal items very light in weight. Each item in a meal was individually packaged in plastic pouches, and all items in the meal were sealed in another pouch. Menus are identified by the day of the mission. NASA recognized that mission activities, such as landing a spacecraft on the lunar surface, would require lots of attention. Meals could be taken whenever it was convenient. Consequently, NASA labeled the meals A, B, and C rather than breakfast, lunch, and dinner (Figure 1.5).

Eating in Microgravity

Children are continually admonished by their parents not to play with their food. Playing with food, though, seems to be a universal human trait. In space, food play literally takes on a new dimension. Astronaut flight videos are filled with amusing sequences showing large spherical drops of fruit punch floating in mid-air, swarms of

M&Ms^{TM}, and Frisbee tortillas sailing across the cabin. You can't blame astronauts for bad manners in space. It's too much fun!

How are astronauts able to do these things? The answer has to do with a term that is not exactly a household word—microgravity. Microgravity is NASA's word for the floating effect that takes place when a spacecraft is in Earth orbit. What exactly is microgravity?

If you watch astronaut videos, you will see that crew members and food items float randomly through the spacecraft cabin. It is easy to get the wrong idea that gravity has gone away. Gravity has not gone away, and the crew and their food are not floating. Instead, they are falling.

The launch of a spacecraft is a battle between the thrust of the rocket and gravity. The amount of energy needed to reach space is enormous. Fortunately, rocket engines can shut off when an orbit is achieved. The spacecraft orbits Earth thousands of times without any additional energy expended. Rather than the absence of gravity, it is gravity that makes orbiting in space possible. If gravity were absent, the spacecraft would travel straight out from Earth and never return.

To see how gravity creates orbits, imagine pitching a baseball horizontally from the top of mountain. To eliminate air resistance, the imaginary mountain has to be tall enough to extend above the atmosphere. Earth's gravity will immediately "grab" onto the ball and cause it arc downward and strike the mountain's flank. Now let's imagine throwing a second pitch faster than before. It arcs, too, but the arc is flatter because of its greater velocity. The ball lands far from the mountain. Keep throwing more balls, increasing the velocity for each throw. Now, Earth's curvature comes into play. The balls not only land successively further away from the foot of the mountain, they begin curving over the horizon before landing. Eventually, one ball makes it completely around Earth, comes back to the mountain-top, and keeps going for another trip. This last ball is traveling so fast that its arc matches the curvature of Earth, and its path is a circle. The baseball is in orbit. The important thing to remember is that while the baseball is orbiting, it is also falling.

When a rocket launches a spacecraft into Earth orbit, it starts by traveling vertically but then pitches forward to travel horizontally. As the rocket increases its velocity, it rises higher. When it climbs above the atmosphere and reaches orbital velocity, the engines shut down. At this point the spacecraft, like the baseball, is falling. Inside, the crew and their food are also falling. This creates the appearance of floating—astronauts and food falling together. Falling is what microgravity is all about.

Microgravity is an environment, created by falling, in which gravity's effects are greatly diminished. Notice that it is gravity's effects and not gravity itself that is diminished by falling.

In microgravity, strange things happen to food. On Earth, gravity helps you measure and pour out a cup of water for cooking. In microgravity, measuring cups don't work very well. If you were to squirt water from a hose into the cup, you would need to do it very slowly. Otherwise, the water would splatter and bounce out in a cloud of hundreds of drops. Once the cup was filled, you wouldn't be able to pour out the water. Liquids are sticky and cling. That's why you have to dry off with a towel after showering. To get the water out of the measuring cup, you would have to shake it, but all of the water would come out in one large drop. Surface tension would pull the water into a shimmering sphere that would gradually dampen into a perfect sphere if it didn't bump into something and splatter (Figure 1.6).

FIGURE 1.6 Astronaut Joe Allen chasing an orange drink on Shuttle. The drink has been squeezed from the package and forms a perfect sphere floating in microgravity (NASA photograph).

Another problem with liquids in microgravity is that sedimentation and buoyancy are absent. Heating water to boiling creates gas bubbles, as it does on Earth, but in microgravity bubbles do not rise to the top of the pot. They stay right where they form. Toss a handful of beans into water and the beans stay at the top. They don't sink.

In short, microgravity creates challenges for food preparation. Then there are the challenges of eating. Whether done on purpose or not, it is easy for food to get away. Too vigorous knife and fork action on a piece of meat can launch it across the cabin.

The potential problems of preparing food and getting it to crew members' mouths were recognized early in the space program. Engineers came to the rescue and designed a variety of systems to help astronauts eat in space.

The different forms of space food and well-balanced menus are only part of the space food story. How they are packaged and eaten is just as important as what kinds of food are packaged.

Space food has to be specially packaged to ensure that the food supply remains safe from contaminants. Space food packages endure more severe environments in space, such as large pressure and temperature changes. Packages must be made to allow for the addition of water, possible mixing, and consumption in microgravity without contaminating the spacecraft environment. Some packages must also withstand heating. In addition to all of these requirements, the packaging material must meet strict NASA specifications with regard to flammability and off-gassing. (Off-gassing is not a problem on Earth. If a package has an unpleasant odor, open a window or turn on the vent fan. In the recirculated atmosphere of a spacecraft, odors are difficult to eliminate, and some can be become downright nauseating.)

In the beginning, food system packaging consisted of tubes and cubes: the dry bite cubes, mentioned earlier, and pureed foods in toothpaste-like tubes. Experience in space led to major improvements. *Apollo* astronauts were provided spoons to consume rehydrated foods from small plastic pouches (Figure 1.7). The menu began to offer much more variety than the foods available in the *Gemini* program. During the three manned missions of *Skylab*, astronauts ate food from open containers with knife, fork, and

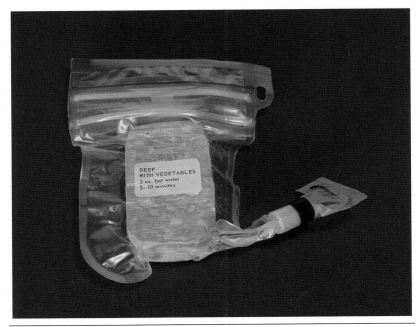

FIGURE 1.7 Apollo Spoonbowl package. The first space food package designed for consumption with a spoon (NASA photograph).

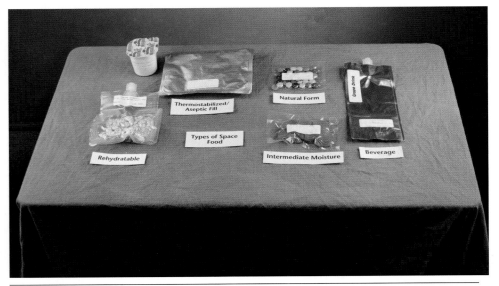

FIGURE 1.8 Shuttle food packages, also used on the ISS (NASA photograph).

spoon. If the food is wet, surface tension will keep the food in the container as long as there are no strong opposing reactions to eject it.

Space shuttle and the ISS crews use several commercial food containers, such as plastic pudding cans and retort pouches (Figure 1.8). Shuttle and ISS beverage packages are modified from Capri Sun® packages made from a foil laminate. A septum, or membrane, allows water to be injected into the beverage package. A large-gauge needle in a wall-mounted galley penetrates the septum. After the right amount of hot or cold water is added and the needle is withdrawn, the septum reseals itself to prevent leakage. A straw is inserted through the septum for drinking. The straw is equipped with a clamp to prevent the liquid from siphoning out when not in use (Figure 1.9).

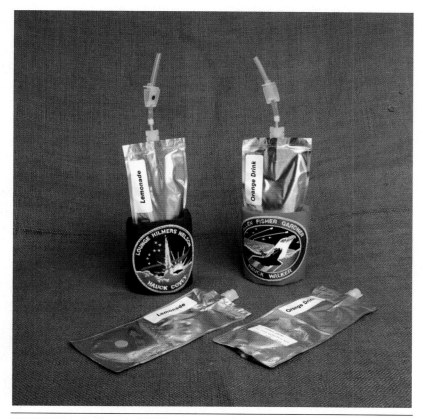

FIGURE 1.9 Shuttle/ISS beverage packages. The experimental beverage holders were only used on a few missions until they were found unnecessary (NASA photograph).

When preparing food for flight, each rehydratable and food package with bite-sized portions is flushed 3 times with nitrogen and sealed at a vacuum of 21–29 in. of Hg. Foods are vacuum packaged to reduce volume, maintain structure, improve rehydration, and reduce oxygen.

Waste generation from space food packaging is always a concern. Storage space is limited on the space shuttle, and all waste is brought back to Earth. An early rigid shuttle food and beverage package was discontinued when the crew sizes increased and waste storage became a problem. Most of the current packages are flexible and easily compressed with a manually operated lever-type compactor.

Become a Space Cook

Each chapter in this book features recipes for spaceflight food. Included are favorite recipes from former astronauts and former NASA dietitians. Some of the ingredients could be difficult to locate because they are mostly used in commercial product preparation or have been discontinued. In this case you can use reasonable substitutes. In some recipes, where the food is normally thermostabilized in retort pouches, a special starch is added. This starch is to aid in the processing and is not needed for home use. Cornstarch is a good substitute. Other starches and special ingredients are listed for some recipes, and substitutes are recommended. These recommendations are the authors' and not NASA's and may not function as well as the original.

Currently, high-dose irradiated products are not available to the general public, and you will not find them in the supermarket. NASA has special permission from the Food and Drug Administration to use high-dose irradiated meats processed under an approved procedure.

As you read the recipes, you will notice the names of some culinary superstars. In 2006 Rachael Ray and Emeril Lagasse separately came to the Johnson Space Center and assisted the Space Food Development Laboratory with the development of more flavorful foods. The recipes developed by these two superstar chefs are included in this book with permission.

The spaceflight recipes in this book have been scaled down from the original recipes. NASA's food specifications are mostly based on 200 servings.

Space food science continually changes with mission requirements, astronaut preferences, commercial availability of the ingredients, and storage considerations. The recipes in this book reflect a certain point in the developing space food system. A few space food recipes were not included due to lack of equipment in most kitchens to process complex items and the availability of special ingredients.

Many people (including some astronauts) wonder why they cannot go to the local supermarket and buy space food. Actually, some foods for space *are* purchased from the supermarket, and others, although they may be found at the supermarket, are purchased directly from the manufacturer. Space food must originate from the same lot for accountability and tracking purposes, and this necessitates purchasing from the manufacturer in many instances. At the end of each chapter is a short list of some space foods that can be purchased from the supermarket.

To Iodize or Not to Iodize, That Is the Question

All of NASA's food specifications call for pure, non-iodized sodium chloride (salt). There is a good scientific reason for this. *Gemini* and *Apollo* spacecraft used chlorine for the spacecraft's water purification system. Chlorine is corrosive, and it created problems with the hardware. When *Skylab* came along, iodine was chosen for the water supply. Storage tanks were treated with iodine before consumption. For the space shuttle, a unique device, called the microbial check valve, was designed. Interestingly, it's not a valve and does not check for microbes. The microbial check valve is a column filled with an iodine resin that imparts iodine to the water as it flows through. Since the water already contains iodine, NASA does not want any more iodine, other than what is naturally in the food, introduced through the food system.

Iodine has both beneficial and harmful effects on human health. Iodine is needed by your thyroid gland to produce thyroid hormones. However, exposure to unnecessarily high levels of iodine can damage the thyroid. It can also affect other parts of your body, such as skin, lung, and reproductive organs.

Because terrestrial water supplies are not normally iodized, it is not necessary to use non-iodized salt in these recipes.

SAUCY COMMENTS

It's been widely reported that John Glenn, Jr., did not like the food NASA sent with him on his *Friendship 7* mission in 1962. Glenn was the first American to orbit Earth. His flight was shortened to four orbits of Earth because of a suspected heat shield malfunction. The problem turned out to be a faulty sensor. Upon Glenn's return, the food story started circulating and is still told today. The news report was faulty, too. Glenn only had time to sample the applesauce, and he thought it tasted pretty good.

AN ENGINEERED SANDWICH

Engineers should probably stay out of the kitchen. *Apollo* meals featured bite-sized cubes of various foods. Following the old candy slogan "Melts in your mouth, not in your hands," the bite cubes could be plucked out of the bag and popped into the mouth, leaving no mess. Take, for example, a chicken sandwich. It would be cut it into cubes, and the cubes freeze dried. Then it would be coated with gelatin to prevent crumbs. Once in your mouth, saliva would rehydrate the sandwiches.

Bite cubes were a brilliant example of spaceflight engineering. Light in weight, compact, easy to use, no mess, and ... yuck! (Figure 1.10).

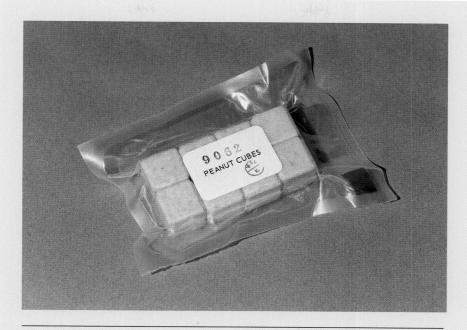

FIGURE 1.10 Apollo peanut cubes, made from peanut butter cookies (NASA photoggraph).

The food had the proper taste. Sugar cookies cubes tasted like sugar cookies, but they didn't look like sugar cookies and they didn't have the sugar cookie texture. In the gastronomic world, this was tantamount to heresy. Bite cubes were discontinued after the *Apollo* program because many of the cubes were found inside the returned space capsule uneaten.

NASA-SPEAK

Like every other NASA program, the Space Food Laboratory has its jargon. Here are some of the "in" space food terms.

Freeze Dehydration. A process whereby prepared food is frozen and dried. This is accomplished by reducing the pressure (vacuum) and thus driving off the water. Freeze dehydration preserves more of the nutrients and flavors than other forms of drying.

Hazard Analysis Critical Control Point (HACCP). A process developed by NASA, The Pillsbury Company, and the U.S. Army Natick Research, Development, and Engineering Center. HACCP involves monitoring and testing critical points in production to insure final quality and is used worldwide by the food and many other industries.

Intermediate Moisture. Dried fruits and meats.

Irradiated. Food that has been zapped with radiation to kill off microorganisms.

Natural Form. "As is" food, such as nuts, cookies, and the ever-present M&Ms.

Rehydration. Adding water to the dry food or beverage before consumption.

Retort Pouch. A foil and plastic pouch containing a pre-prepared entrée. If desired, the pouch can be heated before opening. (These are the classic MREs—meals ready to eat that were first developed for the military.)

Spoonbowl Package. A plastic pouch that permitted the addition of water through a valve and a top that was cut open and the contents consumed with normal utensils.

Thermostabilization. A process of food preservation that heats the food in a closed container to destroy all harmful microorganisms and enzymes. Canning is the non-technical term.

WORST SPACE FOOD?

Many people ask the NASA food staff, "What is the most disliked food?" This is a trick question. The answer is none. If a potential food item is disliked by most of the astronauts, it doesn't get flown on space missions!

MAKE YOUR OWN SPACE FOOD

Many educators ask for instructions on how to make space food. This book provides the answers. Since some of the processing requires equipment and procedures not common to normal kitchens, it is not possible to create and package food that could actually be flown into space. However, fairly authentic-looking replicas can be made if you have a home vacuum sealer. Pick out dry foods or intermediate moisture foods and seal them in packages. Add a contents label and a small Velcro hook patch and the package is ready. Drink packages can be made from dry drink mixes. Measure out the amount needed for one serving and write the amount of water to be added. You will have to squirt the water into the package. In this case, a zip-locking food storage bag might be best. Knead the water and dry mix to encourage proper dissolving. Later, punch the package with a plastic straw cut at an angle and drink. If your food item requires milk, use powdered milk and calculate how much water it will need for rehydration.

Bagging It

Because spacecraft are closed environments, there are many safety restrictions for what can be brought on board. All materials introduced into the spacecraft, including ink in pens and on paper, paper itself, glues, dyes, and labels are subjected to off-gas testing. Materials under consideration for use in space are placed in a low-temperature oven for several hours and monitored for gases produced (bad smells, irritating or toxic fumes, etc.). Materials are even burned to check their flammability and the production of soot.

The cabin atmosphere for the Mercury, Gemini, and Apollo missions was pure oxygen under reduced pressure. However, during a pre-flight simulation for *Apollo 1*, the oxygen atmosphere was at sea level pressure, and materials normally not known for flammability ignited, resulting in the tragic loss of a flight crew. Since then, NASA safety people have paid special attention to potential hazards of everything carried into space. That includes the packages used for food. Not only do they have to pass off-gassing and flammability tests, they also have to be sturdy enough to hold up under reduced cabin pressure. The space shuttle atmosphere is kept at sea level pressure, with normal 78% nitrogen/20% oxygen content. This greatly lowers onboard fire danger. However, the cabin pressure is depressed to 10 psi for several hours to assist astronauts in adjusting to lower pressures as they prepare to go on space walks. This reduces the nitrogen content in their bloodstreams and shortens the time they need to spend in the airlock before going out. Food packages have to be able to withstand this pressure drop without their seals breaking. NASA repackages most space foods in approved packages with known off-gassing and barrier properties.

CHAPTER 2

Breakfast Foods

We've all heard that breakfast is the most important meal of the day. That may be, but for the scientists at the Space Food Laboratory, breakfast is also the most challenging meal of the day. Most typical American breakfast foods, such as eggs over easy, pancakes, bacon, sausage, toast, biscuits and gravy, and various other off-the-grill foods, aren't very good after they have been dehydrated and stored on a shelf for several months.

Breakfast foods not being conducive to dehydration is not the only challenge the food lab folks face. It's also the astronauts. Like the rest of us, when it comes to breakfast, astronauts have many preferences. Some only want black coffee while others demand full-course "lumberjack" breakfasts drowned in syrup.

Scrambled eggs became one of the first breakfast foods of the space program. Cooked, frozen, and freeze dried, the eggs look a bit like tiny bits of yellow Styrofoam pellets. However, when rehydrated, they look, taste, and almost have the texture of fresh scrambled eggs.

One would think scrambled eggs in space would please everybody. Not so! Some astronauts didn't even want to taste them. Their appearance reminded flight crews, many who were still active members of one or another military branch, of the dreaded dehydrated eggs served on base. Those that did try them reported they were messy to eat in microgravity. Little egg bits tended to escape meal packages and drift about until they clung to walls, floors, ceilings, hair, and air vents. In spite of

C.T. Bourland, G.L. Vogt, *The Astronaut's Cookbook*, DOI 10.1007/978-1-4419-0624-3_2,
© Springer Science+Business Media, LLC 2010

their messiness, freeze-dried scrambled eggs have been a space breakfast food staple for decades.

With eggs, astronauts need bacon. Unfortunately, there just isn't a good way to prepare bacon in space. Look at what happens to a kitchen range when frying up a rasher or two. The answer? Bacon bars! Bacon bars were used in the *Gemini*, *Apollo*, and *Skylab* programs. They were made by frying bacon, breaking it into pieces, and compressing it into bars. Bacon bars tasted like bacon, but they lacked the crispy texture.

Breakfast rolls were an immediate success in the labs with flight crews. Just about everybody seemed to like them. You could warm them if you wanted, but otherwise, there was no effort involved other than cutting open the package. Of course, there was a hitch. A typical breakfast roll would stay fresh in the package for seven days—not nearly long enough. The rolls had to be purchased, tested, packaged, and shipped to the Kennedy Space Center weeks ahead of the scheduled liftoff. Considering the complexity of launching a rocket, there was always the possibility of delay. That meant that most of the breakfast rolls that ended up in space were well beyond stale by the time they were consumed.

When the space shuttle came along, a fresh food locker for semi-perishable food was added to the food system. The locker could be stowed at L minus 1 (one day before launch) and swapped with a fresh locker if there was a launch delay. To make the rolls last longer, irradiated rolls were used for the first eight space shuttle missions. These were then replaced thanks to the packaged cake and bread company Sara Lee. Sara Lee began marketing vending machine cinnamon rolls. The flavor of the rolls had to hold up for weeks and weeks in the less than ideal storage environment of vending machines. The new rolls turned out to be perfect for NASA's unique shelf life requirements.

Now, toast was another matter entirely. The options were quite limited. Toast becomes stale very quickly, and the taste goes south. It's also quite crumbly, even fresh from a toaster. Crumbs make a mess and, in microgravity, easily enter nasal passages. For *Gemini* and *Apollo* missions, single bite toasted cubes were created. Nice try! They were more like croutons than toast and were just not popular with the astronauts.

Fresh fruits, another popular breakfast item, create their own problems for space meals. Fruit is carried on the shuttle and the Russian supply ship *Progress* for delivery to the ISS. The quantity of fruit is limited, though, because refrigeration is not available on either of these vehicles. Perhaps future space vehicles will have room for an enriched carbon dioxide atmosphere refrigerator system. On Earth, these systems greatly extend fruit shelf life. For the present, only as much fruit as can be consumed in a few days is carried in space (Figure 2.1)

What does that leave us? Cold cereals, the "breakfast of champions," work pretty well in space. Of course, there are a few choices—with or without sugar and with or without milk. Unfortunately, the milk is powdered. Fresh milk is heavy and doesn't keep.

Just back from space, many astronauts eagerly reach for a cold glass of fresh milk (with cookies, of course). There have been numerous requests to have a tall glass of cold milk in space. NASA and a number of dairy companies worked on the problem, but success has been minimal. Powdered milks either have storage, rehydration, or flavor problems. The best effort to date is commercially produced non-fat-dry milk. It's OK, but it doesn't taste like fresh milk. The ISS

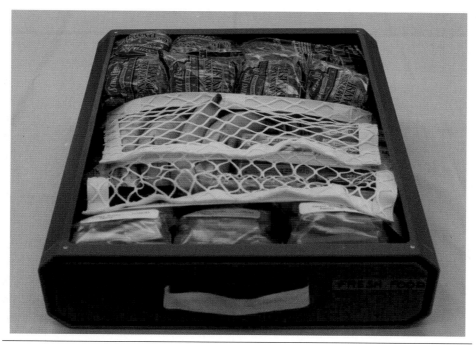

FIGURE 2.1 Shuttle fresh food locker tray with celery, carrots, bananas, and Danish rolls. Note the netting used to hold the food in place in microgravity (NASA photograph).

doesn't have a system to provide cold water for milk rehydration, and the ambient temperature water that has to be used compromises the milk's taste.

That being said, space cereal, milk, and sugar are combined in a rehydratable package. The astronaut injects the correct amount of water and mixes the contents. The cereal has to be eaten immediately, or the contents become soggy. Thankfully, the moisture in a "bowl" of space cereal provides a bit of stickiness that keeps the cereal from slurping out as it being spooned.

SPACE FOOD STICKS

The Pillsbury Company developed a rod-shaped "food stick" to be used inside the *Apollo* space suit. The idea was that it could be inserted through a port in the helmet and into the mouth. The "food stick" was an early attempt to develop a balanced/complete food for emergency use. This eating concept was never fully developed or approved for suit use because the helmet port could not withstand the pressure differentials. NASA did use the space sticks as part of the *Apollo* menu and labeled it "caramel candy." Pillsbury marketed the item as "space food sticks" and later called them "nutrition sticks," but neither strategy increased their sales enough to meet expectations.

WHAT IS THE *PROGRESS CARGO SHIP?*

The *Progress* cargo ship is an unpiloted Russian spacecraft that is launched from the Baikonur Cosmodrome, Kazakhstan, on a *Soyuz* rocket. Among its many duties, such as refueling the ISS's attitude control rockets, it hauls pressurized cargo such as oxygen, food, water, and personal items. A *Progress* spacecraft can carry as much as 7,000 lb of cargo into orbit. As the ISS crew empties a *Progress* of fresh supplies, it is refilled with trash. Finally, it undocks and is sent on its way to burn up in the atmosphere over the Pacific Ocean.

In the recipes that follow, the SS stands for "space shuttle/space station."

SS SCRAMBLED EGGS

5 Grade A large eggs
Egg white from 1 Grade A large egg
1/3 cup 2% milk
1 tbsp nonfat dry milk
2 tsp dried cheddar cheese blend
    (DairiConcepts)*

1 tsp dehydrated cheese seasoning
    (DairiConcepts)*
1/4 tsp salt
1 tbsp unsalted butter

*These are commercial products; you may substitute your favorite cheese powder or grated cheese. Cheese powder is made commercially by spray drying a cheese slurry, using much the same process as when making powdered milk. You may not be able to get these at a local store, but products like them are available on the Internet.

1. Mix whole eggs and egg whites together using a whisk.
2. Combine the milk and nonfat dry milk.
3. Blend the two cheese powders and salt.
4. Using a blender, thoroughly blend the milk mixture and the dry ingredients.
5. Blend milk and cheese mixture with eggs in a saucepan and cook to a semi-coagulated state.
6. Melt butter and blend into the precooked mixture and continue cooking until fully coagulated.

Yield: 6 servings

Note: NASA freeze dries the eggs, adds water back, and freeze dries them a second time. This is necessary in order for the eggs to rehydrate when water is added in the space food package. The basic scrambled egg formulation has been around since the *Apollo* days. Originally a commercial company made them for NASA. When it went out of business the company gave the "secret of the eggs" to NASA to use for space shuttle flights. The "secret" was the second freeze drying. Otherwise, it takes boiling water to rehydrate them, and boiling water is not available on NASA spacecraft.

Variations on the above recipe include:

SS MEXICON SCRAMBLED EGGS

1. Prepare a batch of SS Scrambled Eggs (see recipe above).
2. Add minced fresh green onions, red peppers, and cilantro.

NASA makes its Mexican Scrambled Eggs by adding dehydrated minced green onions, red pepper granules, and freeze-dried cilantro to the freeze-dried eggs to make Mexican scrambled eggs.

SS SEASONED SCRAMBLED EGGS

1. Prepare a batch of SS Scrambled Eggs (see recipe above).
2. Stir in Cugino's Veggie Weggie Dipz™ mix.

NASA adds the dip mix to the freeze-dried scrambled eggs. If you can't find the Cugino's Veggie Weggie Dipz™ mix, try another soup and dip mix, such as Knorr's™. You may have to try several batches to find the correct amount of the soup and dip mix to add to your eggs. That's what the NASA Food Laboratory people do.

BACON BARS

1 lb uncooked bacon

1. Fry the bacon until golden brown.
2. Place the warm bacon into a hamburger press.
3. Exert 3,000 lb of pressure for 10 s.
4. Remove the compressed bacon and let cool.

Yield: More than you would want.

After sampling the bar—so that you could say that you tried it—give the rest to the family dog. One nibble, and Fido will prance about the house barking (Translation: "It's BACON!").

BREAKFAST CEREAL

1 cup of your favorite cold cereal*
1/3 cup of powdered milk
2 tsp of sugar or 1 packet artificial
    sweetener

1/2 cup cold water
1 resealable plastic sandwich bag

*Frosted cereals stay crisper longer than unfrosted cereals.

1. Put all the ingredients in the bag.
2. When ready to eat, add water and reseal the bag.
3. Shake the bag to dissolve the milk and sugar.
4. Open the bag and eat immediately with a spoon.
5. Write a note to yourself to never do that again unless you become an astronaut.

Yield: 1 serving

What You'll Find at Your Supermarket

These breakfast foods eaten by astronauts may be found in the local supermarket. The NASA Food Laboratory staff repackages them in single-serving-sizes in rehydratable pouches.

Multi-Bran Chex™ by General Mills

Kelloggs Frosted Corn Flakes™

Granola or Granola with Raisins™ by Heartland Brands

Mountain House Granola with Blueberries and Milk™ by Oregon Freeze Dry

Instant White Hominy Grits™ by Quaker Oats 30 g plus Butter Buds™

Instant Oatmeal with Maple and Brown Sugar™ by Quaker Oats

Instant Oatmeal with Raisins and Spice™ by Quaker Oats

Rice Krispies™ or Frosted Rice Krispies™ by Kelloggs

Fully Cooked Original Pork Sausage Pattie™ by Jimmy Dean. These have to be freeze-dried before repackaging.

CHAPTER 3

Snacks and Appetizers

Astronauts are no different than the rest of us. Given a chance, they will graze all day. With hundreds of astronauts and support personnel in the Astronaut Office at NASA's Johnson Space Center, someone is always having a birthday or special event. Astronauts are among the world's most efficient hunters of cake and other snack foods. That shouldn't be surprising. An astronaut's day is packed with meetings, training exercises, proficiency flights in training jets, mission simulations, medical tests, and maintaining physical fitness. There is rarely time to sit down and enjoy meals.

In space, schedules are even more hectic. Every moment of every day is planned and plotted on detailed spreadsheets. Meals are scheduled, but they often go by quickly in order to get back to an experiment, vehicle monitoring, Earth observations, or exercise. For snacks in space, it's grab and go.

As with all space food, convenience and utility are paramount. Fortunately, many appetizers and snack foods can be provided to crews nearly straight from the supermarket. These foods are repacked in vacuum-sealed small, tight-fitting plastic bags. All it takes to get into them is a scissors. When it comes to dining on space food, scissors are just as common to flight crews as knives, forks, and spoons are to Earth-bound people. As expected, the bags have small patches of Velcro™ hooks to

C.T. Bourland, G.L. Vogt, *The Astronaut's Cookbook*, DOI 10.1007/978-1-4419-0624-3_3,
© Springer Science+Business Media, LLC 2010

FIGURE 3.1 Astronaut Robert Cabana displays some M & Ms onboard Shuttle mission STS-88 (NASA photograph).

temporarily stick them to corresponding patches of Velcro wool placed throughout the spacecraft cabin. Without the Velcro, snack packs are free to roam the interior of spacecraft and may be harvested by other snacking crew members or stuck on the grills of return air vents (Figure 3.1)

Satisfying Hungry *Skylab* Astronauts

The tradition of space snacks goes back to the early days of human spaceflight. Following the *Apollo* Moon missions, the third stage of one of the remaining Saturn V rockets was converted into an orbital laboratory called *Skylab*. It was launched into space in 1973, and eventually three crews of three astronauts each spent a total of more than 171 days on board. At the time, the third crew of *Skylab* held the record for the number of days in space—84.

With NASA having little long-term experience in spaceflight, the *Skylab* food system was tightly constrained to meet very specific metabolic requirements. A nutrition experiment was established to gather the data needed for future space missions such as trips to Mars and permanent bases on the Moon. The requirements included specific levels of several minerals and proteins that had to be maintained each day.

Some foods were designed to be non-contributors to the *Skylab* mineral and protein balance. That's NASA-speak for snacks. *Skylab* butter cookies were one of the principal non-contributors and could be eaten without jeopardizing the metabolic experiment. The *Skylab* Butter Cookies became sought-after items and were often used as currency for exchanges among the crew. "I'll see your two cookies and raise you. . ."

Skylab Butter Cookies were baked at the Johnson Space Center by Food Laboratory personnel. On one particular Saturday, when the center cafeteria was closed, Food Lab staff used the kitchen to prepare the cookies. A canning machine was brought in, and a cookie production line was created. In the end, a large stack of *Skylab* Butter Cookies in squat aluminum cans with pull-back lids was assembled.

Cookies have since become a spaceflight tradition. Flight crews arrive at the Kennedy Space Center a few days in advance of launch day and are greeted with fresh baked cookies in the Operations and Services Building, where the crew quarters and dining facilities are located. Food personnel schedule their arrival ahead of the crew and begin baking cookies just before the crew arrives. During the baking, the building is saturated with the aroma of fresh-baked cookies and not only astronauts get to partake. Handfuls of cookies are also consumed by the backup crew and by a couple dozen of the local support staff. The food personnel spend many hours baking cookies to keep up with the demand and build up a supply for the next few days. Almost all astronauts eat the cookies, even though a few have complained "You shouldn't be baking these. They're not healthy," as they munched on a cookie.

On Earth, cookies can be large, but in space, size matters. Actually, it's crumbs that matter. Space vehicles are self-contained environments. The quality of the environment is dependent upon minimizing contaminants. Although the air is filtered and carbon dioxide is scrubbed, the atmosphere gets recycled again and again. Crumbs can gum up the works. Foods that produce a lot of crumbs can be a major problem and a big contributor to "pollution" of the air quality during a mission. For this reason, bite-sized cookies and crackers that can be placed in the mouth all at once are preferred to snacks that take several bites to consume. Several regular-size cookies such as pecan "sandies" were removed from the menu on early shuttle missions when astronauts complained that they generated too many crumbs.

SPACEWALKING SNACKS

Imagine trying to eat and drink without the use of your arms and hands. That is what astronauts do when inside their spacesuit. During both *Apollo* and space shuttle spacewalks, astronauts consumed food bars while working in their spacesuits. The food was called the In-suit food bar, and it fit into a sock-like device mounted just inside the neck ring that connects to the helmet (Figure 3.2). Astronauts ate the bar by bending the head forward and biting into it, pulling it out and biting off a piece. This turned out to be unpopular among the astronauts because the bar would rub against the chin and leave sticky goo that could not be cleaned off until the astronaut removed the suit hours later. When NASA discovered that most In-suit bars were being consumed by the astronauts while still in the airlock prior to going into space, they discontinued them.

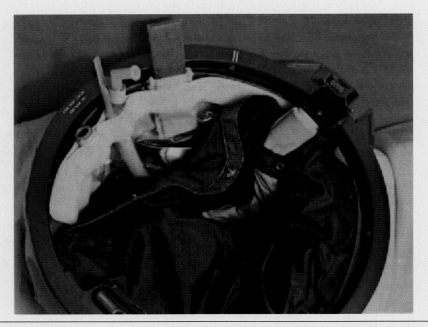

FIGURE 3.2 Neck ring of the Apollo space suit showing the In-suit food bar and the drink device (NASA photograph).

THE MOST POPULAR SPACE SNACK/APPETIZEZR

Shrimp cocktail has long been rated by astronauts as the most popular space food. A study in the 1980s confirmed this for the shuttle by tabulating the number of times it appeared on the menus of astronauts. Shrimp cocktail was chosen more frequently than any other space food. Astronaut Story Musgrave liked his shrimp cocktail so much that he requested it for every meal: breakfast, lunch, and dinner!

Shrimp cocktail has been on NASA space menus since *Apollo* days. Shrimp is an excellent food to demonstrate the freeze dehydration technology. Many foods have significant texture loss when freeze dried. However, when handled properly, shrimp retains most of its texture, and when water is added back it is difficult to tell that the shrimp has been freeze dried.

However, procuring the shrimp and the cocktail sauce has been a problem for NASA. NASA started out buying the freeze-dried shrimp from freeze-drying companies such as Oregon Freeze Dry. Often times this shrimp would not pass the strict microbiological tests required for space food, and at other times it was not available. As a backup, NASA developed a procedure for procuring fresh shrimp and processing them in-house. The secret to passing the microbiological tests was to peel and de-vein the shrimp prior to cooking. The peeling and de-veining after the shrimp were cooked apparently contaminated them with extra microorganisms.

The shrimp cocktail sauce had similar problems. Oregon Freeze Dry made the sauce for commercial sales at one time and then later on made it especially for NASA. It was a proprietary formula, and NASA was never able to duplicate it. When it was not available from Oregon Freeze Dry, a backup source with added horseradish and red pepper was used (Figure 3.3).

FIGURE 3.3 Freeze dried shrimp cocktail with powdered sauce. The current Shuttle flexible package on the left and the original rigid Shuttle rehydratable package on the right (NASA photograph).

CHARLES BOURLAND'S SPACE FOOD DIARY

I was there to meet the first *Skylab* flight crew after their recovery from the ocean by the U.S.S. Ticonderoga. After the longest spaceflight to date, there was great scientific interest in the condition of the crew. As a part of the JSC Food Laboratory, I was there to ensure that the metabolic experiment continued until all medical data had been mined from the crew. That meant that *Skylab* astronauts would have to follow the same metabolic protocol they started 3 weeks before the mission and continue it for 2 weeks after landing. It consisted of providing the astronauts with the same food they had been eating in-flight.

Joe Kerwin, a medical doctor and astronaut on the *Skylab* crew was nauseated when he was transferred from the recovery helicopter to the deck of the Ticonderoga. Joe had been fine during liftoff and 28 days in space onboard *Skylab*. However, reentry and splashdown in the ocean was something else. Stuffed into a bulky *Apollo* spacesuit and bobbing around in the ocean in a tight *Apollo* capsule would tax the strongest stomach.

The medical staff gave Joe some kind of liquid to drink as part of an experiment. He drank it, but his stomach had had enough abuse. The medical guy dropped to the deck and soaked up the emesis with a sponge so that it could be determined how much of the liquid Joe had actually consumed. Talk about unpleasant jobs!

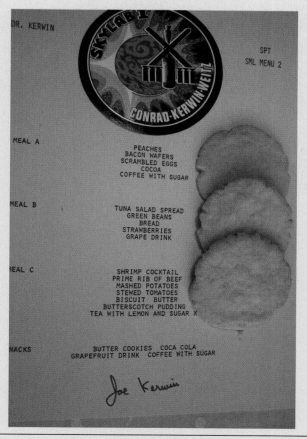

FIGURE 3.4 One day *Skylab* post flight menu for Astronaut Joe Kerwin with some butter cookies.

A little while later, Joe was feeling better. He asked me to give him a butter cookie. Prepared for any eventuality, I had one ready. Joe took a bite and then just released it. The cookie fell to the deck and broke. Joe said "I guess I will have to get used to gravity again" (Figure 3.4).

SKYLAB BUTTER COOKIES
(served natural form)

6 tbsp unsalted butter   1¼ tsp vanilla
4 tbsp sugar
4½ tbsp maltodextrin*   1 cup plus 2 tbsp cake flour

*NASA used Maltrin 100 from the Grain Processing Co.; however, it is only available in 100-lb packages. Maltodextrin may be purchased in health food stores and is usually called complex carbohydrate powder. It may also be substituted for with sugar.

1. Allow butter to come to room temperature. Using a mixer, cream the butter with the sugar and the maltodextrin.
2. Sift the remaining ingredients together and stir them into the mix.
3. Roll into small balls about 3/4 in. in diameter. Place on baking sheet and flatten.
4. Bake 15 minutes at 325°F. Let cool.

Yield: 18 cookies

SS SHRIMP COCKTAIL

Shrimp:

4 lb Individually Quick Frozen large (approx. 25–35 per pound) peeled and de-veined shrimp or 4 lb fresh large peeled, de-headed and de-veined Gulf shrimp

Shrimp Boil Mixture:

1 bag dry crab boil
4 tbsp bottled lemon juice
1 tbsp dehydrated onion flakes
4 tsp Tabasco sauce

1 tbsp celery salt
1 tbsp garlic powder
1 tsp salt

1. Rinse shrimp thoroughly with water and soak for 10 minutes in 1.5% salt solution (3 tbsp of salt per gallon of water).
2. Drain the shrimp and heat 1 gal of water.
3. Add the shrimp boil mixture to 1 gal of water and heat to boiling.
4. Add shrimp to the boiling mixture and boil for 6–8 minutes. Drain immediately and chill with ice or place in refrigerator.
5. Serve chilled.

Since dried shrimp cocktail sauce is usually not available on the retail level the best substitute is a store-bought sauce. Add some extra horseradish to give the space shrimp cocktail a real kick.

Yield: 16 servings

Note: For space NASA freeze dries the shrimp and adds dried cocktail sauce to the shrimp at the time of packaging. In orbit, astronauts merely add chilled water to the package and dissolve the sauce by kneading the package.

JAN DAVIS'S HOT CLAM-CHEESE DIP

2 small onions, chopped
1 4-oz can chopped green chilies
6 tbsp butter
2 10½ oz cans minced clams,
   drained
1 lb processed cheese, cut into
   cubes

8 tbsp catsup
2 tbsp Worcestershire sauce
2 tbsp sherry or milk
½ tsp cayenne pepper

1. Sauté onions and chilies in butter.
2. Add remaining ingredients. Cook until cheese melts.
3. Serve hot with dipping-style corn chips.

Yield: 8 servings

Meet the Astronaut: **Jan Davis, STS-47, STS-60, STS-85**

Dr. Davis received undergraduate degrees in Applied Biology from Georgia Institute of Technology and in Mechanical Engineering from Auburn University in 1975 and 1977, respectively. She got her Master of Science degree in 1983 and a doctorate in Mechanical Engineering from the University of Alabama in Huntsville in 1985. Dr. Davis became an astronaut in June 1987. A veteran of three spaceflights, she has logged over 673 hours in space. She flew as a mission specialist in 1992 and 1994 and was the payload commander in 1997. Dr. Davis has the distinct honor of having her flight picture on the cover of the NASA book *Nutrition in Spaceflight and Weightless Models* (CRC 2000).

LEROY CHIAO'S CHINESE COLD PEANUT NOODLES

1/2 lb dry vermicelli noodles
2 cloves garlic, chopped
2 tbsp peanut butter
1 tsp sesame oil

Chinese hot (spicy) oil to taste
1/4 cup peanuts, diced
1/4 cup chopped green onions
    (scallions)

1. Boil noodles to desired consistency, then drain and rinse with cold water.
2. Mix noodles with garlic, peanut butter, sesame oil, and spicy oil to taste.
3. Sprinkle top with diced peanuts and green onions. Serve chilled.

Yield: 4 servings

Meet the Astronaut: Leroy Chiao, STS-65, STS-72, STS-92, and Expedition 10 (6½ months aboard ISS)

Dr. Chiao received a Bachelor of Science degree in Chemical Engineering from the University of California, Berkeley, in 1983, and a Master of Science degree and a doctorate in Chemical Engineering from the University of California, Santa Barbara, in 1985 and 1987, respectively. He was selected by NASA in January 1990 and became an astronaut in July 1991. A veteran of four spaceflights, he flew as a Mission Specialist on STS-65 (July 8–23, 1994), STS-72 (January 11–20, 1996) and STS-92 (October 11–24, 2000), and was the Commander and NASA Science Officer on Expedition-10 (October 13–April 24, 2005). Dr. Chiao has logged a total of 229 days, 7 hours, 38 minutes and 5 seconds in space, including 36 hours and 7 minutes of EVA time in six spacewalks.

GRACE NELSON'S HOT CRAB DIP

1 lb fresh or canned Florida crab
   meat
1 cup mayonnaise

1 cup Parmesan cheese
Toast points

1. Mix ingredients, then warm in a double boiler.
2. Serve toast points on the side.
3. Serve in a warm chafing dish.

Meet the Astronaut: **Bill Nelson, STS-61C**

Bill Nelson, the husband of Grace Nelson, is a graduate of Yale University and the University of Virginia Law School. In 1972, he was elected to the Florida legislature, where he served for 6 years until he was elected to Congress in 1978. He was the second U.S. senator to go into space (Senator Jake Garn was the first). STS-61C *Columbia* (January 12–18, 1986) launched from the Kennedy Space Center, Florida, and returned to a night landing at Edwards Air Force Base in California. During the six-day flight the seven-man crew aboard *Columbia* deployed the SATCOM KU satellite and conducted experiments in astrophysics and materials processing. The space experience gave Senator Nelson, who has always been an advocate of space, a better appreciation of what is involved in sending people into space.

IN-SUIT FOOD BAR

Fruit leather*　　　　　　　　　Wax paper
Water　　　　　　　　　　　　Q-tip

*The brand known as Fruit Roll-ups™ are a type of fruit leather. If you happen to have a food dehydrater, you should follow the instructions that came with it for making your own fruit leather.

1. Spread out and stack layers of fruit leather to ¼ in. depth.
2. Using a Q-tip, slightly dampen the leather between layers.
3. Cover the layers with wax paper and place a weight on the leather for a couple of hours. A brick or a heavy skillet will do.
4. Remove the wax paper and cut into strips 1 in. wide by 9 in. long.

Note: Astronaut In-Suit Bars were covered with edible film, but since this is not readily available you can coat yours with wax paper and remove before consuming. NASA started out making several flavors but later on combined all the flavors into one multi-flavor bar to reduce inventory requirements.

What You'll Find at Your Supermarket

Snacks and appetizers are among the easiest of space foods to obtain on Earth. Just go to the supermarket and check the cracker, nut, cereal, and cookie aisles to find a wide selection. Here are some of the choices.

> Applesauce in small disposable containers (SOPACKO produces carbohydrate-enhanced applesauce in retort pouches)
> Chessmen Butter Cookies™ by Pepperidge Farms
> Chocolate-coated Almonds™ by Masterfoods
> M&M Plain Chocolate Candies™ by Masterfoods
> M&M Chocolate Peanuts™ by Masterfoods
> Almonds, roasted and salted macadamia nuts, and toasted and salted cashews

Cheddar Cheese spread-Squeezers™ (individual serving packets are made by Portion Pac Inc.)
Toasted wheat crackers
Dried apricots
Fruit cocktail in cans
Granola bars
Canned peaches
Peanut Butter-Squeezers™ (creamy peanut butter in individual packets are produced by Portion Pac, Inc.)
Dry roasted peanuts
Lorna Doone™ shortbread cookies by Kraft

CHAPTER 4

Soups and Salads

A meal is not a meal unless there's soup. That's the Russian space program's concept for space lunches and dinners. NASA has always included a few soups on the menu to provide space crews with some extra menu variety. However, when NASA's food specialists began working with the Russian food specialists to develop menus for missions to the *Mir* space station, soups became a priority.

In the 1990s, important agreements for American/ Russian cooperation in space were signed. The agreements allowed for astronauts and cosmonauts to fly on each other's space vehicles and for Americans to work on the *Mir* space station. This cooperation ultimately led to the creation of the International Space Station (ISS), which routinely includes Americans and Russians as a part of the crew.

As a result of this new cooperation in space, feeding crews became more complicated. Menus had to reflect cultural differences. Soup was important to the cosmonauts, and so NASA expanded its soup offerings. The mixed crews were given the opportunity to sample each other's delicacies, which sometimes stretched the limits of diplomacy.

There are just some items you have to grow up with in order to appreciate. American astronauts were hard pressed to consume Russian borscht, a bright red soup made from beetroot. The Russians looked askance at the great American

C.T. Bourland, G.L. Vogt, *The Astronaut's Cookbook*, DOI 10.1007/978-1-4419-0624-3_4,
© Springer Science+Business Media, LLC 2010

culinary institution, peanut butter. Aleksei Leonov, the first person to take a space walk and one of two Russian members on the 1975 *Apollo-Soyuz* mission, gagged on a sample and tried to spit it out.

One area of universal agreement was fresh fruit and vegetables. Both have been "must have" menu items since the early days of the space shuttle. Part of the reason for this was that two new categories of astronauts were created for the shuttle program—Mission Specialists and Payload Specialists. These were scientists and engineers selected to conduct research in space and perform other space tasks such as deploying payloads and assembling structures. Often, Mission and Payload Specialists came from the ranks of the young and health-conscious. To satisfy them, fresh fruits and vegetables (apples, bananas, and carrot and celery sticks) began to fly and have been a part of the food manifest ever since. Occasionally, oranges, pears, nectarines, grapefruit, and, for the stout of stomach, jalapeno peppers have flown (Figures 4.1 and 4.2).

FIGURE 4.1 Japanese astronaut Mamoru Mohri with Japanese apple on Shuttle. This particular variety of apple is highly recognizable in Japan, but banned in the US, so NASA had to get special permission from the USDA to import them and follow a strict protocol to insure all seeds were destroyed. (NASA photograph).

FIGURE 4.2 Astronaut Rhea Seddon eating an orange from a meal tray on an early Shuttle mission. Note the peels attached to the rubber holders on the meal tray. (NASA photograph).

Some of the challenges of providing fresh fruit and salads to flight crews were discussed in the snack and appetizer chapter. There's more!

All fresh fruits and veggies have to be consumed in the first 2–3 days of the mission. One of the problems with fruit is that it is metabolically active and expels odors while ripening. Stowed in the spacecraft several hours before launch, the odors accumulate in the closed environment. Some astronauts find the smells objectionable, especially if they become nauseated when entering microgravity (and about half of them do). It would seem like the solution is to seal the fruit and veggies in air-tight bags. Unfortunately, that speeds up the metabolic activity, and spoilage is accelerated. Because of the odor problem, bananas and oranges have lost some of their popularity with crews. Some space shuttle commanders, whose word is final, have been known to order "no bananas on my flight."

When it comes to cosmonauts, the most popular fresh foods are onions and garlic. These are always included on the *Progress* re-supply ships when food is delivered to space. In spite of their popularity, these items tend to have a divisive effect on the crew.

TWINKIES® IN SPACE!

When American astronauts began flying on board the Russian *Mir* space station, there was more to be concerned about than just conducting experiments in space. Due to language difficulties and limited communications with home, Norm Thagard, the first American to work on *Mir*, felt isolated. After 140 days in space, Shannon Lucid replaced him. To learn from experience, the psychology folks at the Johnson Space Center decided to boost morale and send her a care package on a *Progress* resupply spacecraft. In the package they put books, records, and comfort foods, including Twinkies! On its way to space, the package had to be approved in Russia. The Russians refused to ship the Twinkies because the package did not have an expiration date. The Food Lab joked that the package didn't have to have an expiration date because Twinkies never expire. In the end, the Twinkies did not make the journey to *Mir*.

JELLO AGAIN!

Having comfort foods in space works wonders. According to Shannon Lucid, "It is the greatest improvement in spaceflight since my first flight over ten years ago. When I found out that there was a refrigerator on board *Mir*, I asked the food folks at JSC if they could put Jello™ in a drink bag. Once aboard *Mir*, we could just add hot water, put the bag in the refrigerator, and, later, have a great treat. Well, the food folks did just that and sent a variety of flavors for me to try out. We tried the Jello™ first as a special treat for Easter. It was so great that we decided the *Mir-21-NASA 2* crew tradition would be to share a bag of Jello™ every Sunday night. (Every once in a while, Yuri will come up to me and say, "Isn't today Sunday?" and I will say "No, it's not. No Jello™ tonight!!!")" (Figure 4.3).

FIGURE 4.3 Cosmonauts Yury Onufyyenko, Yury Usachev and Astronaut Shannon Lucid share a meal on the Russian Mir Space Station. (NASA photograph)

SS CHICKEN NOODLE SOUP (FIGURE 4.4)

1/2 cup dried fettuccine noodles

3 tbsp National 150 filling aid starch from the National Starch and Chemical Co. (may substitute with 1 tbsp cornstarch)

1 tbsp soft white wheat dark toasted flour from Breiss Malt and Ingredient Co.

1/2 tsp salt

1/2 tsp coarse ground black pepper

1/4 tsp dried parsley flakes

1/4 tsp poultry seasoning

11/2 cup water

11/4 cup College Inn Frozen 16% Concentrated Chicken Broth™

Available in the northeast United States or on the Internet. May substitute with your favorite concentrated chicken broth

1/4 cup half and half (milk and cream)

11/4 cup Individually Quick Frozen 3/4 in. diced natural proportion chicken meat #674314 from Valley Fresh. NASA uses frozen, but you may substitute with fresh.

1/2 cup sliced carrots (1/2 in. piece)

1/4 cup diced celery (1/2 in. pieces)

1/4 cup fresh diced yellow onions (1/2 in. pieces)

1. Break fettuccine to decrease length and blanch in boiling water for 5 minutes and drain.
2. Combine starch, flour, salt, black pepper, parsley, and poultry seasoning and mix thoroughly.
3. Add starch mixture to ½ cup water, mix well, and set aside.
4. Add remaining water and broth to a saucepan and mix. Heat mixture to simmer.
5. Stir in the starch mixture and half and half.
6. Add chicken, carrots, celery, and onion to soup and mix well.
7. Add the fettuccine and heat to boiling. Simmer until chicken and carrots are tender.

Yield: 6 servings

Note: NASA further processes the Chicken Noodle Soup by thermo-processing it in a retort pouch.

FIGURE 4.4 Chicken noodle soup.

SS CITRUS SALAD

1 24-oz jar citrus salad in extra light
syrup

1 24-oz jar Mandarin oranges in
light syrup

1. Drain citrus salad and Mandarin oranges separately in colanders for approximately 5 minutes.
2. Combine 7 oz of the drained Mandarin oranges with the drained citrus salad and carefully mix well so as not to break up the fruit pieces.
3. Chill and serve.

Yield: 6 servings

Note: NASA further processes the Citrus Salad by thermo-processing it in a retort pouch.

SS CREAM OF MUSHROOM SOUP

2 tbsp unsalted butter
4 tbsp canned chopped
  mushrooms, stems and pieces
  included
2 tbsp Minor's Mushroom Base™.
  Available on the Internet or
  substitute with another brand

1/4 cup white bleached all-purpose
  flour
2 $\frac{1}{3}$ cups water
1½ cups half and half

1. Melt butter in a large saucepan.
2. Puree canned mushroom pieces.
3. Warm mushroom base and half and half together in small pot.
4. Blend flour into butter using a whisk. Stir over medium heat for 7–9 minutes until mixture is bubbly and well blended. Turn off heat.
5. Gradually add water to the flour and mix well.
6. Add mushroom base and pureed mushrooms to the water-flour mix and heat to boiling. Boil and stir for 1 minute.
7. Add half and half, stir well, and serve.

Yield: 6 servings

Note: NASA freeze dries the Cream of Mushroom Soup prior to use. It purees the mushrooms so the soup can be consumed through the beverage straw in microgravity, but this step is not necessary if consumed with a spoon.

SS PEACH AMBROSIA (FIGURE 4.5)

4 tbsp pecans
1¾ cups peaches, fresh or frozen
1/2 cup fresh Bartlett pears or
   canned pear halves in water and
   juice concentrate
1 ¾ cups canned, sliced pineapple

1 medium fresh banana
1 tsp erythorbic acid (see below for
   substitute suggestion)
1 tsp salt

1. Chop pecans into small pieces.
2. Dice peaches and pears into 1/2-in. pieces.
3. Drain pineapple and cut slices into pieces that are approximately ½ in. at the widest point.
4. Peel bananas and dice into 1/2-in. pieces. To prevent browning of banana during the peeling and dicing stage, place banana in a solution of 1 tsp erythorbic acid and 1 tsp salt and place in 1quart of water. After dicing rinse bananas with water. Lemon juice may be used to prevent browning if erythorbic acid is not available.
5. Mix peaches, pears, pineapple, bananas, and pecans. Chill and serve.

FIGURE 4.5 Peach Ambrosia.

Yield: 6 servings

Note: NASA further processes the Peach Ambrosia by freeze drying prior to packaging. The pecans are freeze dried separately and added to the freeze dried fruit.

Peach Ambrosia has been a part of space food programs for many years, including on menus for *Apollo, Skylab,* the shuttle, and the ISS.

SS SPLIT PEA SOUP

1.5 oz Block & Barrel, Imperial, Endless Hickory hearth ham or your favorite ham
1 cup dry split green peas
2¾ cups water
1 tsp ham base with no added MSG and with smoke flavoring added (#14-203) from Eaten Foods Co. This is a food service product, but substitutes are available on the Internet
1/4 tsp coarse ground black pepper
2 tbsp whole milk
1 tbsp National 150 filling aid starch from National Starch and Chemical Co. (may substitute with cornstarch)

1. Remove skin from ham and dice into ¼ in. pieces.
2. Rinse peas and sort through to discard any foreign matter.
3. Combine water, ham, split peas, black pepper, and ham base in a saucepan.
4. Heat on medium, simmer, and stir frequently until peas have disintegrated (approximately 1½ hours).
5. Add starch to milk to make a slurry; mix well, and add to the cooked peas.
6. Heat another 5 minutes and serve.

Yield: 6 servings

Note: NASA further processes the Split Pea Soup by thermo-processing it in a retort pouch.

SS POTATO SOUP

2¾ cups water

1 tsp Butter Buds 8X™ from Butter Buds Ingredients. You can substitute with 2 tbsp regular Butter Buds™

2 tbsp National 150 filling aid starch from National Starch and Chemical Co. (may use cornstarch as a substitute)

½ tsp ground black pepper

½ tsp salt

1 cup Sysco Classic Potato Pearls, Excel Instant Mashed Potatoes™ distributed by Sysco Corporation. You may substitute with your favorite instant mashed potatoes

1 tbsp unsalted butter

4 tbsp onions, diced into ¼ in. pieces

1¼ tsp bottled minced garlic in water

1 tsp vegetarian Vegetable Base #14-403 from Eatem Foods. You can substitute with your favorite vegetable base available at gourmet food stores or from the Internet

1½ cups whole milk

1¼ cups peeled ½ in. diced Natural potatoes distributed by Sysco Corporation. You can use instead fresh red potatoes

1 tsp Clearjel modified food starch from the National Starch and Chemical Co. (Optional)

Pinch of freeze-dried chives

1. Combine starch, Butter Buds, black pepper, salt, and ¼ cup water. Mix well and set aside.
2. Combine instant potato pearls and 1½ cups water. Mix well and set aside.
3. Add 1 cup water, butter, onions, garlic, and vegetable base to a saucepan and begin heating.
4. Heat until the butter melts, then add the potato pearl mixture and mix well.
5. Stir the starch mixture and add to the pan.
6. Heat to simmer and hold for 3–5 minutes.
7. Add the milk, chives, and potatoes and heat to boiling.
8. Simmer until potatoes are tender.

Yield: serves 6

Note: NASA further processes Potato Soup by thermo-processing in a retort pouch.

SS TOMATO BASIL SOUP (FIGURE 4.6)

14-oz can diced tomatoes in juice

10-oz can crushed tomatoes

1/4 tsp vegetable base, Mirepoix, no MSG from Eatem Foods Co. A food service product, substitute with your favorite vegetable base available at gourmet food stores or the Internet.

1/8 tsp caramelized garlic base from Eatem Foods Co. You may substitute with your favorite vegetable base available at gourmet food stores or the Internet.

1/2 cup skim milk

Dash of salt

Dash dried oregano

Dash coarse ground black pepper

2 tsp dried whole basil leaves

3 tbsp heavy whipping cream

1. Add crushed tomatoes, diced tomatoes, vegetable base, and caramelized garlic base in a saucepan. Mix well and heat over a medium flame.
2. Add milk, salt, pepper, oregano, and basil to the mixture and mix well. Continue heating.
3. Add the cream and mix well. Heat to boiling and turn down flame, simmering 3–5 minutes.

Yield: 6 servings

Note: NASA further processes Tomato Basil Soup by thermo-processing it in a retort pouch.

FIGURE 4.6 Tomato basil soup.

SS RHUBARB APPLESAUCE

6 oz frozen sliced strawberries
12 oz frozen rhubarb,
   approximately ½ in. thick stalks.
   If stalks are longer than 1½ in.,
   cut or break into 1½ in.. NASA

uses frozen, but you may
   substitute with fresh
   12-oz canned or bottled
   unsweetened applesauce
Sugar to taste

1. Thaw strawberries and blend in a blender to a smooth puree.
2. Combine pureed strawberries, applesauce, and rhubarb and mix well.
3. Heat in a saucepan on medium for 15 minutes or until the rhubarb is tender.
4. Add sugar as needed. Chill and serve.

Astronauts eat this at room temperature because there is no refrigerator on the shuttle or on the ISS.

Yield: 6 servings

Note: NASA further processes Rhubarb Applesauce by thermo-processing it in a retort pouch.

SS TROPICAL FRUIT SALAD

1 24-oz jar mango in extra light
   syrup

1 24-oz jar tropical mixed fruit in
   light syrup with passion fruit
   juice

1. Drain mango in a colander for approximately 5 minutes and cut into ½ in. cubes.
2. Drain tropical mixed fruit for approximately 5 minutes.
3. Add about half of the cubed mango to the drained tropical mixed fruit and carefully mix well so as not to break up the fruit pieces.
4. Chill and serve.

Yield: 6 servings

Note: NASA further processes the Tropical Fruit Salad by thermo-processing it in a retort pouch.

PAULA HALL'S* HILL COUNTRY POTATO SALAD

Dressing:

1/3 cup vegetable oil
1/3 cup cider vinegar
1/4 cup fresh lemon juice
4 pickled jalapenos (CAUTION: This will be hot!)
4 cloves of garlic

1 tsp ground cumin
1 tsp dried oregano
1 tsp freshly ground pepper
1 tsp salt

Salad:

6 red potatoes, unpeeled, cut into quarters
1 red onion, sliced
1/4 cup fresh cilantro, chopped

1 cup whole kernel corn, canned or frozen
2 red bell peppers, cut into julienne strips
6 green onions, chopped

1. Combine all ingredients for dressing in a blender container. Process at high speed until smooth.
2. Cook the potatoes in boiling water in a saucepan until tender; drain.
3. Toss with the dressing in a large bowl.
4. Add the onion, cilantro, corn, red peppers, and green onions, tossing to coat.
5. Serve chilled or at room temperature.

Yield: 12 servings

*Former shuttle/ISS dietitian for 10 years, who lost her battle with cancer in 2007.

BILL POGUE'S VINEGAR SLAW

1 head cabbage (can use half green and half red cabbage; this makes the slaw pink)

Sauce:

1 cup white vinegar   1 cup water
1 cup sugar   1/8 tsp Tabasco Sauce

1. Cut cabbage into 2-in. chunks. Put some into blender and cover with water.
2. Pulse gently on CHOP. Drain through strainer and pour into bowl. Continue until all is chopped.
3. Mix sauce ingredients and set aside until sugar is dissolved. Stir and mix with cabbage (it should cover cabbage).
4. Refrigerate overnight.

Yield: 10–12 servings

Meet the Astronaut: **Bill Pogue,** *Skylab 4* **(84 days)**

Colonel Pogue received his Bachelor of Science degree in Education from Oklahoma Baptist University in 1951 and Master's degree in Mathematics from Oklahoma State University in 1960. He enlisted in the air force in 1951 and received his commission in 1952. While serving with the Fifth Air Force during the Korean Conflict, from 1953 to 1954, he completed a combat tour in fighter bombers. From 1955 to 1957, he was a member of the USAF Thunderbirds. Colonel Pogue is one of nineteen astronauts selected by NASA in April 1966. He was the pilot of *Skylab 4* (third and final manned visit to the *Skylab* orbital workshop), launched November 16, 1973, and concluded February 8, 1974. This was the longest manned flight (84 days, 1 hour and 15 minutes) in the history of US manned space exploration at that time.

CONNIE STADLER'S* STRAWBERRY RHUBARB SALAD

*Former *Apollo*, *Skylab*, and shuttle dietitian from 1970 to 1988.

4 cups rhubarb
2 tbsp water
1 cup sugar
1 package strawberry-flavored
   gelatin

1-lb package frozen strawberries
or 1 pint fresh strawberries,
mashed

1. Cut rhubarb into 1-in. pieces.
2. Place in baking dish and sprinkle with water and sugar.
3. Bake in 350°F oven for 40 minutes.
4. Remove from oven and stir in the dry gelatin immediately.
5. Chill until the mixture starts to set, then add the strawberries and chill for several more hours until firm.

What You'll Find at Your Supermarket

Chicken Consommé (Minors vegetarian consommé prep, chicken style, by Nestle)

CHAPTER 5

Bread, Tortillas, and Sandwiches

Bread, one of the oldest and most popular of all commercially prepared foods, is a big problem for spaceflight. As mentioned before, there is the crumb problem. The crumbs go everywhere in microgravity. But keeping bread fresh and tasty enough for eating is the real challenge.

During the *Apollo* Moon flights, the crew cabin door had to be opened occasionally for spacewalks. The door did not have an airlock. All three crew members had to don their spacesuits in preparation for a single crew member to go outside. Cabin air was bled away, and the hatch was then opened.

After the spacewalk, the cabin was sealed and flooded with fresh oxygen. Suits were removed and stowed. The astronauts were hungry, and individual slices of bread were on the menu. With vacuum conditions inside the cabin during the spacewalk, air inside the bread packages began to expand and pop the seal. All air inside the packages leaked out. This released the nitrogen gas packed with the bread to retard spoilage. The bread was still OK to eat, at least at first, but the loss of nitrogen and the leak in the package shortened the shelf life of bread slices reserved for future meals. It was suggested that the problem of popping "bread bags" would be solved by vacuum-sealing bread prior to flight. All that strategy accomplished was to pre-squish the bread into a kind of "bread leather."

C.T. Bourland, G.L. Vogt, *The Astronaut's Cookbook*, DOI 10.1007/978-1-4419-0624-3_5,
© Springer Science+Business Media, LLC 2010

One novel solution for the bread was to can it. During the *Skylab* mission, bread was actually baked inside small cans and sealed. The idea worked, but it provided only a single dinner roll per can. Another strategy included irradiating the bread to kill microorganisms and prolong freshness. Timing was everything, and even with irradiation, the flight stowage process took so long that the bread was well past freshness before liftoff.

Frozen sandwiches, made to crew member specifications, are placed in flight suit pockets or carried onboard in some other way for first meals in space. This practice began after an astronaut on the *Gemini 3* mission (usually said to be John Young) smuggled a corned beef sandwich in his flight suit. The astronaut got "busted" after word leaked out. NASA was concerned about potential safety problems, and reporters had a fun story to pursue.

One interesting discovery about bread in space was made. Individual slices of rye bread had the best shelf life of all bread tried for flight. However, the rye bread eventually lost its appeal if used too frequently.

Single-slice bread packages continued to be flown on the first space shuttle missions. The shuttle cabin has an airlock, and when spacewalks were scheduled, the cabin air did not have to be bled away before the door could be opened. The popping bread package problem was solved. There were still the crumbs and the staleness problem. Astronaut Mary Cleave and Payload Specialist Rodolfo Vela (the first Mexican astronaut) introduced tortillas to the shuttle menu in 1985 (Figure 5.1). It was almost a "duh" moment for spaceflight. Tortillas produce very few crumbs. They can be rolled up into one-handed sandwiches and make great cabin Frisbees. Tortillas were an immediate hit and have flown ever since.

The first fresh tortillas were obtained from a local Houston tortilla factory and hand carried to the Kennedy Space Center. When shuttle missions were extended to longer periods in space (up to 17 days), freshness became a problem. With no refrigeration on board, the tortillas would often develop mold after 5–7 days.

Since tortillas were the primary bread for the shuttle, food lab researchers began an effort to develop an extended shelf-life tortilla. A similar effort for bread, used in the military for ready

FIGURE 5.1 Mexican payload specialist Rudolfo Vela Neri, on Shuttle with a tortilla. Astronaut Mary Cleave and Rudolfo are credited with introducing tortillas on Shuttle (NASA photograph).

to eat meals, had achieved success. The concept employed reduced water activity and depleted oxygen. Water activity is a measurement of the water available for microorganisms to multiply. Reducing the water activity prevents the growth of microorganisms. Formulation changes and replacing some water with glycerin achieved the reduction. To cut back on oxygen, the tortillas were packaged in a foil pouch flushed with nitrogen. Furthermore, an oxygen scavenger packet was placed inside the pouch to remove remaining oxygen in the tortilla. Mold cannot grow without oxygen (Figure 5.2).

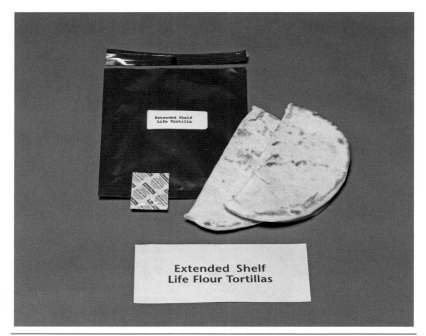

FIGURE 5.2 Extended shelf life tortillas. The package is sealed in a nitrogen atmosphere and the oxygen scavenger packet removes any residual oxygen (NASA photograph).

NASA made extended shelf life tortillas for several years. One of the drawbacks with the extended shelf life tortillas was that they became bitter tasting after 6 months in storage. When Taco Bell came out with the extended shelf life tortilla in the late 1990s NASA tested them. They found that the Taco Bell tortillas would store for 12 months without any bitter flavor development. NASA began using the Taco Bell tortillas for extended duration missions like those on the ISS. However, they still use fresh tortillas for short shuttle missions.

Not satisfied with just tortillas, NASA has conducted research on bread making in space. Astronauts on multi-year missions to Mars will likely have to grow some of their own food. A high-yield wheat has been developed specifically for spaceflight by scientists at the University of Utah. The wheat has short stalks and can grow under continuous lighting conditions (Figure 5.3). Such a plan will require a small flour mill

FIGURE 5.3 Test subject Nigel Packam inside a sealed chamber that used wheat to produce his oxygen (NASA photograph).

and an oven for baking. NASA has communicated with a Swedish company working on a prototype bread machine that might work in microgravity. Unfortunately, the electrical power requirements for the bread maker far exceeded what is available on ISS. Perhaps, future astronauts will have a bakery in the permanent base they establish on the Moon.

The combinations for space sandwiches in space are endless. Astronauts like to spread cheese, tuna, and chicken salad, bean dip, peanut butter and jelly, and many meats on their tortillas (Figure 5.4). There is a skill to making tortilla roll-ups in space. You have to be conservative in the amounts of ingredients you put in them. If you think eating tortilla roll-ups is messy on Earth...

FIGURE 5.4 Astronaut Franklin Chang-Diaz spreads bean dip on a tortilla (NASA photograph).

FIRST SANDWICH IN SPACE

Mercury astronauts didn't think much of the few foods they were permitted to eat in space. Gus Grissom, pilot of the second *Mercury* mission, liked the food less than anybody and wasn't shy about saying so. Gus was assigned to the *Gemini 3* mission along with new astronaut John Young. Young got Wally Schirra, another *Mercury* astronaut, to pick up a corned beef sandwich from Wolfies Deli in Cocoa Beach. In orbit, when Gus was supposed to have a meal, John pulled out the sandwich from a flight suit pocket and said, "Here skipper, want to try a sandwich?" Gus took a couple of bites and realized the bread was too crumbly. He stopped eating it.

The *Gemini 3* crew successfully demonstrated that manned spacecraft could be maneuvered in space. However, word of the sandwich got out, and *Gemini 3* is now remembered as the corned beef sandwich flight. The corned beef sandwich was even discussed in congressional budget hearings. Deke Slayton, head of the Astronaut Office, had to put an official reprimand in John's personnel file. The reprimand didn't hurt John's future as an astronaut, though. He eventually walked on the Moon and commanded the first space shuttle flight. Sadly, Grissom lost his life in the tragic *Apollo 1* fire along with Roger Chaffee and Ed White.

DEADLY TUNA

On one of the *Apollo* missions, an astronaut decided to eat a can of tuna salad left open from the day before. Since there wasn't any refrigeration on board, one of the other of the crew said to him, "You're gonna get sick. You may die." The worried astronaut called down from space and asked if it was OK to eat it. He was told no with no uncertainty: "Absolutely, definitely do not eat that. Open up another one if you want to eat tuna salad." Fortunately, the tuna salad was a fairly acidic product, and nothing happened to the astronaut from having eaten it. Being an astronaut with an iron stomach probably helped, too.

SPACE FOOD TRIVIA

What do you think are the five most popular foods eaten in space?

1. Shrimp Cocktail 2. Lemonade 3. Steak 4. M&Ms® 5. Brownies

FROZEN SPACE SANDWICH RECIPE

2 slices white, whole wheat, or rye
    bread, with no crumbly coatings
Lunch meat (your choice)

Cheese (your choice)
Condiment packets (your choice)
Sealable plastic bag

1. Place meat and cheese on bread.
2. Slice diagonally.
3. Place in bag, seal, and freeze.
4. Remove from freezer and thaw at room temperature.
5. Add condiments and eat within 6 hours or discard.

Yield: 1 serving

TORTILLA ROLL-UP

1 tortilla
Peanut butter, refried beans,
    chicken salad, or anything else

in your pantry that strikes your
    fancy
Jelly (optional)

Spread the filling on the tortilla and roll it up.

Caution: If you are in microgravity and plan on using your tortilla as a Frisbee, save the filling step for later. Unevenly spread filling can affect the flight handling characteristics of the tortilla Frisbee and may lead to scattered globs of filling or tortilla Frisbee art adhering to cabin walls.

What You'll Find at Your Supermarket

6-in. flour tortillas (soft taco kits only)
Whole wheat flat bread
Dinner rolls
Extended shelf life waffles by DeWafelbakkers, Inc.
Cinnamon Raisin Danish by Sara Lee
Little Debbie™ snack fudge brownies
Prepackaged cinnamon rolls

CHAPTER 6

Main Dishes

What's for dinner? That's a question all of us ask just about every day. Sometimes, it is hard enough just to plan the evening meal in advance, much less having to think about meals for the rest of the week. Debating daily dinner selections was not something *Gemini* or *Apollo* astronauts did during their flights. Their dinner menus were selected for them 6 months in advance!

At first, everybody ate the same thing. Their meals were prepackaged. The appetizer, entrée, vegetable, condiments, drink, and dessert were contained in a single package. If you didn't feel like a meal combination on a particular day, you were pretty much out of luck.

Of course, astronauts could beat the system. If the meal package contained chocolate pudding and you really wanted cookies, you could open another day's meal and swap items. It meant a little busywork to make sure the raided meal package was resealed and stowed. Then, there was the temptation of having two desserts one day and none the next—feast or famine.

In the beginning of the US manned space program, astronauts were pretty interchangeable (although none would admit it). They could be no taller than 5'11" and weigh no more than 185 lb. They had to have jet test pilot experience and a tolerance for bizarre medical tests. Although American women were not

C.T. Bourland, G.L. Vogt, *The Astronaut's Cookbook*, DOI 10.1007/978-1-4419-0624-3_6,
© Springer Science+Business Media, LLC 2010

prohibited from spaceflight, they didn't get the chance to prove their equality until more than 20 years after the first human traveled into space.

The first cracks in the "right stuff" astronaut program came during the *Apollo* program. Harrison Schmidt, a geologist, got to walk on the Moon during the *Apollo 17* mission. Then came medical doctor Joseph Kerwin as part of the three-man crew for the *Skylab 2* mission. Jack Lousma, on the *Skylab 3* mission, broke the mold in a different way. He was the tallest and heaviest astronaut to fly in space at that time. As a result of his extra size, his caloric requirements were greater than his crewmates. He got to double up on some entrées, like two cans of spaghetti and meatballs for dinner.

The desire of astronauts to select their meals at mealtime rather than have it decided for them 6 months ahead became a "perfect storm" on one space shuttle mission. The crew, insisting on having it their way, spent valuable orbital time at the beginning of their mission unpacking all of their food items and re-stowing them pantry style. They proved that a pantry system was feasible for food storage on the shuttle, and it made the crew happier. Naturally there was resistance to this from the medical folks, nutritionists, and dietitians who wanted to monitor what and how much astronauts were consuming as a part of a comprehensive health-monitoring program. After more than two decades of their health being monitored in space, the astronauts rebelled. As one astronaut put it, "I am 40 years old and have made it this far without a dietitian or doctor watching what I eat."

Space menus are still picked 6 months in advance. Breakfasts, lunches, and dinners for the entire crew are stowed in separate lockers, or individual lockers hold complete mission meals for each astronaut.

Foods served on the ISS are still based on the individual meal system. This helps the food laboratory staff members determine how much food should be sent into space for the crew. Once onboard the ISS, the food is stored in a pantry. Crew members are permitted to indulge themselves but are expected to demonstrate common sense in their dietary choices.

DO ASTRONAUTS HAVE GOOD TASTE?

It has always been widely suspected that microgravity has a significant effect on the ability of astronauts to taste food in space. Without Earth's gravitational effects, normal convection currents are not present. The aromas from a fragrant bowl of hot soup will not rise to the nose to register a sensation in the brain.

Whether or not taste changes in space is debatable. Some astronauts say there is a definite change, while others report no difference. Two in-flight and one ground-based experiment failed to settle this issue. Theoretically, there should be some difference in taste, either due to not being able to smell the food as they normally do or to physiological changes that occur in the human body. In microgravity, body fluids tend to accumulate in the upper torso, resulting in congestion.

CHARLES BOURLAND'S DIARY: A BOOTLEG OPERATION

When our first shipment of space food was shipped to Russia, Russian customs officials sent word that NASA would have to pay a large sum of money to process it through their customs. Additionally, Russian customs insisted on breaking the seals on the containers for inspection. That caused a big problem, because Russian food specialists would not accept containers with broken seals. It became "Catch 22" time. Finally, we had the shipment returned to us, as a deadline was looming. Luckily, a group of NASA specialists were heading to Russia for a meeting. Each of us was given a box of space food to check through as baggage. The Russian customs office was circumvented. After the incident, an agreement was reached to bypass Russian customs in the future.

CHARLES BOURLAND'S DIARY: GEORGIA BBQ

When astronaut Sonny Carter from Georgia was assigned to the STS-33 shuttle mission, he immediately began campaigning to add a favorite pork BBQ to the menu. I told him to bring back a sample the next time he was in Georgia. We freeze dried it, did a taste panel evaluation, and then sent it to the microbiological laboratory for analysis. The taste panel results were good, but the sample failed the microbiological tests in several categories. I told Sonny that the BBQ failed microbiological tests and consequently could not be used in space. I thought that would be the end of it but should have known better. People like Sonny don't get to be astronauts by accepting "no" for an answer. An MD, pro soccer player, and fighter pilot, Sonny was ready for Round Two. He thought the sample might have been contaminated in transit and brought in another for testing. It failed as well. Sonny kept at it, and I finally asked him to give me the telephone number of the BBQ producer. I learned that to produce the BBQ, the pork shoulders were roasted and the meat was hand pulled. I asked the producer to cook a shoulder and remove it from the oven with sterile gloves, package and freeze it, and ship it frozen to Houston. We pulled the meat from the bone using aseptic procedures, and it passed the microbiological tests. Sonny was finally rewarded for his efforts and got to dine on Georgia BBQ in space. In spite of the fuss, the Georgia BBQ was not popular enough with other crews to continue it on the shuttle food list. (Sadly, Sonny became another astronaut to lose his life during the program. He was killed in a commercial plane crash while on NASA business.)

SS SLICED BEEF WITH BBQ SAUCE (FIGURE 6.1)

2 lb beef round

2¼ cups honey BBQ sauce

1 tbsp apple cider vinegar

1. Cook beef in a 350°F oven to medium well done (showing 150°F using a meat thermometer).
2. Cool, trim, and slice into ¼ in. slabs.
3. Mix vinegar and BBQ sauce.
4. Place beef and BBQ sauce mix in a baking dish and reheat.

Yield: 4 servings

Note: NASA further processes the Sliced Beef with BBQ Sauce by thermo-processing it in a retort pouch.

FIGURE 6.1 Sliced beef with BBQ sauce.

SS CHICKEN WITH CORN AND BLACK BEANS

1 lb Southwest Medley, Individually Quick Frozen corn/black bean medley from Jon-Lin Foods, Colton CA. Food service product; substitute with fresh or frozen corn and black beans.

6 oz chicken Individually Quick Frozen, Classic ½-in. fully cooked white meat from Sysco Corporation. Food service product; NASA uses frozen, but you can substitute with fresh.

8 oz canned diced tomatoes in puree

1/4 tsp garlic powder

1/8 tsp salt

Pinch ground black pepper

1. Place all ingredients in a saucepan and bring to boil.
2. Reduce heat and simmer until beans and corn are tender.

Yield: 6 servings

Note: NASA further processes Chicken with Corn and Black Beans by thermo-processing it in a retort pouch.

SS CHICKEN SALAD

2 lb skinned chicken breasts

2 tbsp melted butter

4 sticks celery

1 medium size red onion

1 cup plus 4 tbsp low-fat or fat-free mayonnaise

1. Preheat oven to 350°F.
2. Place chicken on trays and baste both sides with melted butter.
3. Bake to an internal temperature of 190°F, approximately 20 minutes
4. Flip chicken, baste with butter, rotate pan, and bake an additional 20 minutes.
5. Remove chicken from oven, allow to cool slightly, and dice into approximately ¼ in. cubes.
6. Dice celery and red onions into approximately ¼ in. pieces
7. Combine diced chicken with the mayonnaise, celery, and red onion. Mix well, chill, and serve.

Yield: 7 servings

Note: NASA further processes the Chicken Salad by freeze drying.

SS CORN BREAD DRESSING (FIGURE 6.2)

The cornbread is an ingredient and has to be made first.

Cornbread:

2/3 cup skim milk

1/3 cup Egg Beaters™ or other egg substitute

1 cup yellow corn meal

1 cup unbleached all-purpose flour

1 tbsp extra fine sugar

2½ tsp baking powder

1. Grease an 8 × 8 in. baking pan with butter or margarine.
2. Place milk and egg product into a bowl and mix well.
3. Add cornmeal, flour, sugar, and baking powder. Mix until thoroughly combined.
4. Pour into baking pan and bake approximately 12 minutes at 350°F or until golden brown and a toothpick inserted in center comes out clean.
5. Cool cornbread to room temperature and crumble into coarse crumbs. Put in a mixing bowl and set aside.

FIGURE 6.2 Corn bread dressing.

Cornbread Dressing:

1/4 medium yellow onion
1 medium stalk celery
1 tbsp unsalted butter
1/4 tsp salt
1 tsp dried poultry seasoning

1/4 tsp coarse ground black pepper
1/4 tsp dried parsley flakes
1/4 tsp dried rubbed sage
1 cup reduced sodium chicken broth

1. Preheat oven to 325°F.
2. Grease an 8 × 8 in. baking pan with butter or margarine.
3. Peel onion, puree in a food processor, and set aside.
4. Trim ends of celery and finely chop in a food processor. Add to onion puree.
5. Heat sauté pan over medium heat. Melt butter and sauté onion puree and celery until celery is soft. Add to crumbled cornbread.
6. In a separate bowl combine salt, poultry seasoning, black pepper, parsley, and sage. Add to cornbread-sautéed vegetable mixture and mix well.
7. Pour chicken broth over cornbread mixture and mix well.
8. Spoon dressing into the baking pan.
9. Bake at 325°F for 35 minutes.
10. Remove from oven and serve.

Yield: 6 servings

Note: NASA further processes the Cornbread Dressing by freeze drying.

SS MEATLOAF (FIGURE 6.3)

2 tbsp onion soup mix
2 tbsp saltine-style crackers,
   coarsely crushed
1/4 tsp coarse ground black pepper
1/8 tsp dried oregano leaves
1/8 tsp dried basil leaves
1 tsp minced garlic

2 tbsp plus 1 tsp Egg Beaters™ egg
   substitute
2 lb lean ground beef
1 cup canned crushed tomatoes
1/3 cup canned or bottled chili sauce
2 tbsp light brown sugar
1/8 tsp distilled white vinegar

1. Combine the onion soup mix, crackers, black pepper, oregano, and basil and mix well.
2. Add the garlic and egg product and mix well.
3. Blend in the ground beef and tomatoes (NASA uses a ribbon blender and refrigerates the mixture overnight.)
4. Preheat convection oven to 450°F. NASA uses a convection oven, but a conventional oven will work, too.
5. Line two baking sheets with aluminum foil and spray with nonstick vegetable cooking spray.

FIGURE 6.3 Meatloaf.

6. Scoop out half-cup quantities of the meatloaf mixture, form into oval shaped patties, and place on the baking sheets.
7. Cook the meatloaf patties for approximately 15 minutes (convection oven) (18 minutes conventional) or until browned and cooked through. Internal cooking temperature should be 170–185°F.
8. Combine the chili sauce, brown sugar, and vinegar and heat.
9. Serve the meatloaf and pour the sauce over it.

Yield: 10 servings

Note: NASA further processes the Meatloaf by adding the meatloaf and a portion of the sauce to a retort pouch and thermo-processing.

SS SWEET 'N SOUR CHICKEN

3 lb boneless and skinless chicken breasts
1/3 cup Sweet & Sour Sauce (NASA uses the dried mix, but the liquid is more readily available and equally suitable.)

1 tbsp Clearjel instant starch from national Starch and Chemical Co. (may substitute cornstarch)
3 tbsp salted butter

1. Preheat oven to 350°F.
2. Place chicken on trays and baste both sides with melted butter.
3. Bake to an internal temperature of 190°F, approximately 20 minutes; flip chicken, baste with butter, rotate pan, and bake an additional 20 minutes if needed.
4. Remove chicken from oven, allow to cool slightly, and dice into approximately 1/2 in. cubes.
5. Mix the sweet and sour sauce and the instant starch and add water to rehydrate.
6. Mix with the chicken cubes and heat and serve.

Yield: 6 servings

Note: NASA further processes the Sweet 'n Sour Chicken by freeze drying the chicken and adding the sauce and starch.

SS CHICKEN-PINEAPPLE SALAD (FIGURE 6.4)

2 lb deboned and skinned chicken breasts

3 tbsp salted butter

1 1/3 cups drained canned pineapple tidbits in their own juice

3/4 cup plus 2 tbsp nonfat mayonnaise dressing

1/2 cup diced celery pieces, about 1/4 in. in size

1/3 cup chopped pecans

1. Preheat oven to 350°F.
2. Place chicken on trays and baste both sides with melted butter.
3. Bake to an internal temperature of 190°F for about 20 minutes. Flip chicken, baste with butter, rotate pan, and bake an additional 20 minutes or until 190°F is reached.
4. Remove chicken from oven, allow to cool slightly, and dice into approximately 1/4 in. cubes.
5. Mix chicken, pineapple, dressing, celery, and pecans and chill before serving.

Yield: 6 servings

Note: NASA further processes the Chicken-Pineapple Salad by freeze drying.

FIGURE 6.4 Chicken-pineapple salad.

SS RED BEANS AND RICE (FIGURE 6.5)

11 oz dried small red beans

5½ cups distilled water

1/2 cup Individually Quick Frozen parboiled rice # QF-P-00000-64 from Sage V Foods. Fully cooked food service product; can substitute with cooked rice

3 tbsp diced celery

3 tbsp diced onions

3¼ cups water

Pinch garlic powder

Pinch ground cayenne pepper

Pinch ground dried oregano

1/8 tsp salt

2 tbsp cornstarch

1½ tsp Tabasco™ sauce

1 tbsp caramelized garlic base # 99-404 from Eatem Foods. Food service product; can substitute with chopped garlic in water

1 tbsp caramelized onion base # 99-425 from Eatem Foods. Food service product; can substitute with onion powder

1 1/3 cup navy bean flakes from Inland Empire Foods. Fully cooked dried food service product; can use mashed navy beans as a substitute

1/2 tsp Liquid Smoke™

FIGURE 6.5 Red beans and rice.

1. Sort through the red beans and discard any foreign parts. Rinse with cold tap water.
2. Add distilled water and allow the beans to soak overnight.
3. Remove beans and rinse well in tap water. Drain well.
4. Combine soaked beans, rice, celery, and onions. Set aside.
5. Combine garlic powder, cayenne pepper, oregano, salt, and starch with ¼ cup of water. Mix well and set aside.
6. Add remaining water to bean mixture followed by the Tabasco™, caramelized garlic base, and caramelized onion base. Mix well.
7. Add bean flakes to pot, mix well, and let stand for 30 minutes.
8. Heat mixture to simmer and add the starch mixture while stirring.
9. Add Liquid Smoke to beans and rice mixture and continue heating to boiling. Simmer until beans are tender.

Yield: serves 6

Note: NASA further processes the Red Beans and Rice by thermo-processing in a retort pouch.

RACHAEL RAY'S 5 VEGETABLE FRIED RICE WITH 5-SPICE PORK

2 cups chicken stock
3/4 cup water
1½ cups cooked white rice
5 tbsp vegetable oil, divided
1 lb thin-cut pork loin chops
Salt
Pepper
2 tsp Chinese 5-spice powder
2 eggs, beaten
1/2 lb shitake mushrooms, thinly sliced

1/2 cup carrots, shredded
1 red bell pepper, seeded, quartered lengthwise, and cut into 1/4 in. slices
1 scallion, thinly sliced on an angle
1 cup green peas, fresh or frozen
3 cloves garlic, finely chopped
2 in. fresh ginger root, grated or minced
1/2 cup dark soy sauce (Tamari)

1. Bring stock and water to a boil. Add rice, stir, cover, and cook for about 18 minutes, then fluff with a fork and turn out onto a cookie sheet to cool.
2. Just before taking the rice off the stove, heat a deep nonstick skillet or wok over high heat with 2 tbsp of vegetable oil.
3. Thinly slice the pork and season with salt, pepper, and 2 teaspoons of 5-spice powder. Stir fry the meat, then push off to the sides or transfer to a holding plate.
4. Add another tablespoon of vegetable oil and heat. Then add eggs and scramble, breaking into small bits. Push eggs off to side of pan.
5. Add remaining 2 tbsp of vegetable oil to pan. Heat oil, then stir fry the mushrooms, carrots, and red peppers for 2 minutes. Add scallions, peas, garlic, and ginger and toss around for one minute more.
6. Add rice and sauté mixture for a couple of minutes, then douse with Tamari and mix in the pork.

Yield: 4 servings

Note: When used in space NASA further processes the 5 Vegetable Fried Rice with 5-Spice Pork by freeze drying.

RACHAEL RAY'S SWEDISH MEATBALLS (FIGURE 6.6)

1/3 lb ground beef
1/3 lb ground pork
1/3 lb ground veal
1 egg
1/2 cup plain bread crumbs
1/4 cup cream
3 tbsp finely chopped white onion
1/4 tsp dried mustard
1/8 tsp grated nutmeg
Salt
Pepper
2 tbsp unsalted butter

2 tbsp white, unbleached flour
2 cups beef stock
1 cup sour cream
2 tsp of lingonberry preserves, red currant jelly, or grape jelly
1 lb egg noodles, cooked al dente
finely chopped fresh dill, for garnish
finely chopped parsley, for garnish
1/2 cup finely chopped cornichons or baby gherkin pickles, optional

1. Preheat oven to 400°F.
2. In a bowl, mix meats with egg, bread crumbs, ¼ cup cream, chopped onions, dried mustard, nutmeg, salt, and pepper.
3. Roll the mixture into 1-in. balls and arrange on a nonstick baking sheet.

FIGURE 6.6 Rachael Ray's Swedish Meatballs.

4. Bake 10–12 minutes.
5. Heat a saucepot over medium heat. Melt the butter, whisk in flour, and cook for 1–2 minutes. Whisk in beef stock and thicken 6–8 minutes.
6. Stir in sour cream and jelly and warm through.
7. Season finished sauce with salt and pepper, to taste.
8. Remove balls carefully with a thin spatula. Mix meatballs and egg noodles with sauce and garnish with dill, parsley, and, if you wish, finely chopped cornichons or baby gherkin pickles.

Yield: 4 servings

Note: When used in space NASA further processes the Swedish Meatballs by freeze drying.

RACHAEL RAY'S MINI FLORENTINE TURKEY MEATBALLS WITH ORZO (FIGURE 6.7)

1 10 oz box frozen spinach, defrosted in microwave
1 lb ground turkey breast
3 tbsp finely chopped white onion
2 cloves garlic, finely chopped
1 egg
1½ cups milk, divided
1/2 cup bread crumbs, a couple of generous handfuls
1/2 cup grated Parmesan cheese
Coarse salt
Black pepper

Olive oil, for generous drizzling
2 tbsp butter
2 tbsp flour
1 cup chicken stock
2 cups shredded provolone cheese (1 10-oz sack)
1/4 tsp grated nutmeg, approximately
Handful of finely chopped flat leaf parsley
1/2 lb Orzo pasta, cooked al dente

1. Preheat oven to 425°F.
2. Wring the spinach completely dry using a kitchen towel.
3. Place turkey in a bowl and combine with spinach, 3 tbsp of finely chopped onions, the garlic, the egg, a splash of the milk, the bread crumbs, and the grated cheese.

FIGURE 6.7 Rachael Ray's Mini Florentine Turkey meatballs with Orzo.

4. Season the turkey with salt and pepper and add a generous drizzle of olive oil to the bowl. Mix the meat and roll into small 1-in. balls. Arrange on a nonstick baking sheet lightly prepared with oil or cooking spray.
5. Bake balls 10–12 minutes until juices run clear.
6. In a medium saucepot over medium heat melt butter, then whisk in flour and cook a minute or two more.
7. Whisk in the milk and the stock. Stir until sauce thickens, 4–5 minutes.
8. Season with nutmeg, salt, and pepper, then melt in provolone and Parmesan cheese. Adjust seasonings to taste.
9. Use a spatula to release meatballs and combine with sauce and pasta. Sprinkle with parsley and serve.

Yield: 4 servings

Note: When used in space NASA further processes the Mini Florentine Turkey Meatballs with Orzo by freeze drying.

RACHAEL RAY'S TACO CHILI MAC

2 tbsp corn oil (two turns of the pan)
2 lb ground sirloin
2 jalapeno peppers, seeded and
    chopped
1 onion, finely chopped
4 cloves garlic, finely chopped
Salt
Pepper
3 tbsp chili powder, a couple of
    healthy palmfuls
1 rounded tablespoon cumin, a
    healthy palmful

1 bottle of beer or 1½ cups beef
    broth
1 28-oz can crushed fire roasted
    tomatoes
1 lb small pasta: mini rigatoni with
    lines, corkscrews, or elbows
Shredded smoked or sharp
    cheddar, for garnish
Chopped green olives with
    pimientos, for garnish

1. Heat a deep skillet over medium high heat. Add oil and then meat.
2. Brown meat, then add peppers, onions, and garlic. Season with salt and pepper, chili, and cumin, and cook until tender, 6–8 minutes.
3. Deglaze the pan with beer or broth and stir in tomatoes. Heat through.
4. Mix tomato sauce with pasta and top with cheese and olives.

Yield: 6 servings

Note: When used in space NASA further processes the Taco Chili Mac by freeze drying.

EMERIL'S KICKED UP BACON CHEESE MASHED POTATOES

4 baking potatoes, such as russets (about 3 lb), peeled and cut into 1-in. pieces
1¾ tsp salt
1/2 cup heavy cream
4 tbsp butter
1/4 tsp ground black pepper

8 slices bacon, cooked crisp and crumbled
1/2 lb sharp cheddar cheese, grated
1/4 cup sour cream
1/4 cup chopped fresh chives
Freshly ground black pepper

1. Place the potatoes and 1 teaspoon of salt in a heavy 4-quart saucepan. Add enough water to cover the potatoes by 1 in. Bring to a boil.
2. Reduce the heat to a simmer and cook until the potatoes are fork tender, about 20 minutes. Cooking time will be less for smaller portion sizes.
3. Drain in a colander and return potatoes to the cooking pot. Add the cream, butter, remaining ¾ teaspoon salt, and black pepper.
4. Place the pan over medium low heat and mash with a potato masher until you've achieved a light texture, about 4–5 minutes.
5. Add the bacon, grated cheese, sour cream, and chopped chives and stir until thoroughly combined.

Yield: 4–6 servings

Note: When used in space NASA further processes the Kicked Up Bacon Cheese Mashed Potatoes by freeze drying.

EMERIL'S MARDI GRAS JAMBALAYA

1 5-lb duck, trimmed of fat and cut
   into 8 pieces
3 tbsp Emeril's Original Essence™
2 tbsp vegetable oil
1 lb Andouille or other spicy
   smoked sausage, diced
2 cups chopped onions
1/2 cup chopped green bell
   peppers
1/2 cup red bell peppers, chopped
1/2 cup chopped celery
1 tsp salt, or more to taste
1/2 tsp cayenne pepper
1/2 tsp freshly ground black pepper

2 cups peeled, seeded, and
   chopped tomatoes
1 tbsp garlic, chopped
3 bay leaves
2 cups uncooked long-grain white rice
2 tsp fresh thyme, minced
2 quarts chicken stock or canned
   low-sodium chicken broth
1 lb small shrimp, peeled and
   deveined
1 cup chopped green onions (green
   and white parts)
1/2 cup minced fresh flat-leaf
   parsley

1. Season the duck pieces with 2 tbsp of the Essence. If you don't have Essence, use your favorite Creole seasoning.
2. Heat the vegetable oil in a large heavy pot over medium-high heat. Add the duck, skin side down, and sear for 5 minutes.
3. Turn and sear on the second side for 3 minutes. Remove from the pot and drain on paper towels.
4. Add the sausage to the fat in the pot and cook, stirring, until browned, about 5 minutes.
5. Add the onions, bell peppers, celery, salt, cayenne, and black pepper and cook, stirring often, until the vegetables are softened, about 5 minutes.
6. Add the tomatoes, garlic, and bay leaves and cook, stirring, until the tomatoes give off some of their juices, about 2 minutes. Add the rice and cook, stirring, for 2 minutes.
7. Add the thyme, chicken stock, and duck. Bring to a boil. Reduce the heat to medium-low, cover, and simmer, stirring occasionally, until the rice is tender, about 30 minutes.
8. Remove duck pieces from the jambalaya and cool slightly.
9. Discard skin and bones and shred duck meat. Return the duck meat to the rice mixture.
10. Season the shrimp with the remaining Essence. Add the shrimp to the pot and cook until they turn pink, about 5 minutes.

11. Remove the pot from the heat and let sit, covered, for 15 minutes.
12. Add the green onions and parsley to the jambalaya and stir gently. Remove and discard the bay leaves.
13. Adjust the salt, pepper, and cayenne to taste.

Yield: 6 servings

Note: When used in space NASA further processes the Mardi Gras Jambalaya by freeze drying.

RACHAEL RAY'S SPICY THAI CHICKEN WITH RED PEPPERS AND BASIL

1½ cups jasmine rice, prepared to package directions
1 tbsp light vegetable or peanut oil (one turn of the pan)
1 tbsp hot chili oil (one turn of the pan)
1½ lb thin cut chicken breast
1 onion, thinly sliced
2 red bell peppers, seeded and very thinly sliced

4 cloves garlic, finely chopped
1 tsp coarse black pepper
1/4 cup dark soy (Tamari)
A few dashes Thai fish sauce, about 1 tsp
2 cups (about 40 leaves) fresh basil
1/4 cup salted peanuts, chopped
A handful of cilantro, finely chopped

1. Start rice according to package directions.
2. Heat oils in large nonstick skillet or wok over high heat.
3. Shred chicken into thin strips and cut into bite-sized pieces. Add chicken and stir fry until golden, 2–3 minutes.
4. Push chicken off to the sides of the skillet and add the onions and peppers to the center of the pan. Stir fry 2–3 minutes, then combine with meat and add the garlic and pepper. Stir 1 minute, then add soy and fish sauce; adjust seasonings to taste, tear basil into pieces and wilt in.
5. Remove from heat and serve over rice. Garnish with chopped salted peanuts and cilantro.

Yield: 4 servings

Note: When used in space NASA further processes the Spicy Thai Chicken with Red Peppers and Basil by freeze drying.

PAULA HALL'S CHIPOTLE-LIME MARINATED GRILLED PORK CHOPS

4 boneless or bone-in chops, about
   1¼ in. thick
1 chipotle chili, canned in adobo,
   chopped, OR 1 dried chipotle
   chili, rehydrated and minced
2 tsp oregano

2 garlic cloves, crushed
2 tbsp vegetable oil
2/3 cup lime juice
1 tbsp cilantro, chopped
1/2 tsp salt

1. Place chops in a large self-sealing plastic bag; combine remaining ingredients in a small bowl and pour over chops.
2. Seal bag and refrigerate for 4–24 hours.
3. Remove chops from marinade (discard marinade) and grill over medium-hot coals for a total of 12–15 minutes, turning to brown evenly. Serve chops immediately.

Yield: 4 servings

GLORIA MONGAN'S* FAJITAS

1/4 cup lime juice (approx. 2 large
   limes)
1 tsp salt
1/2 tsp garlic powder
1/8 tsp cayenne pepper
1/4 tsp black pepper
1 lb flank steak, trimmed and
   scored
4 chicken breasts, boneless and
   skinless
1 medium red onion, thinly sliced

1 red bell pepper, cut into thin
   strips
1 green bell pepper, cut into thin
   strips
1 tsp margarine
4 6-in. flour tortillas, warmed in
   oven
1/4 cup chopped tomatoes
1/4 cup lettuce, shredded
1/4 cup picante sauce
2 tbsp reduced-calorie sour cream

*Mongan was a shuttle and ISS dietitian from 1988 to 2004.

1. Mix lime juice, salt, garlic powder, cayenne, and black pepper in large shallow dish. Add steak and chicken; turn to coat.
2. Refrigerate covered, 4 hours or overnight, turning once.

3. Remove steak and chicken from marinade. Broil or grill 6 in. from heat, turning once until desired doneness (about 8–10 minutes for medium rare).
4. Saute onion and peppers in margarine until soft.
5. Slice steak and chicken across grain into thin strips.
6. Divide among tortillas and serve with tomatoes, lettuce, picante sauce, and sour cream.

Yield: 4 servings

GLORIA MONGAN'S KAHLUA GRILLED SHRIMP ON ANGEL HAIR PASTA

3 lb peeled and cleaned shrimp

Marinade:

1 cup Kahlua (a coffee flavored liqueur)
1½ cups honey
1½ cups vegetable oil
2 10-ounce bottles Tiger Sauce™
2 tbsp seasoned salt

2 tbsp garlic, chopped
1 tbsp parsley, chopped
2 ounces hot pepper sauce
1 tbsp fresh basil, chopped
1 tbsp fresh thyme, chopped
2 tbsp fresh cilantro, chopped

Pasta:

2 tbsp Worcestershire sauce
4 cups beef broth

1 tbsp red pepper flakes
1 lb cooked angel hair pasta

1. Combine ingredients for marinade, reserving 1 cup for sauce. Marinate shrimp for at least 2 hours.
2. Grill shrimp.
3. Heat Worcestershire sauce, beef broth, and red pepper together.
4. Add cooked pasta to beef broth.
5. Serve shrimp over angel hair pasta.

Yield: 9 servings

JOE KERWIN'S SOUR CREAM CHICKEN ENCHILADAS

8 oz sour cream

1 10 3/4-oz can cream of chicken soup

1½ cups chicken broth

1 4-oz can green chilies, chopped

12 flour tortillas

2½–3 lb chicken, cooked, boned, and shredded (reserve broth)

8 oz Monterey Jack cheese, grated

4 oz Cheddar cheese

1. Combine sour cream, soup, broth, and green chilies. Heat and stir until smooth and well blended.
2. To soften tortillas, heat a small amount of reserved chicken broth in a skillet. Place tortillas, one at a time, in broth for a few seconds. Remove and drain.
3. After draining, place tortillas, one at a time, directly into soup mixture.
4. Lift tortillas out of the soup mixture. Place 3 tbsp chicken and 2–3 tbsp of each cheese in the center of each tortilla.

Meet the Astronaut: **Joe Kerwin, Skylab 2 (28 days)**

Dr. Kerwin received a bachelor's degree in Philosophy from the College of the Holy Cross, Worcester, Massachusetts, in 1953, and a doctor of Medicine degree from Northwestern University Medical School, Chicago, Illinois, in 1957. He attended the US Navy School of Aviation Medicine in Pensacola, Florida, being designated a naval flight surgeon in December 1958. He earned his wings at Beeville, Texas, in 1962 and was selected as a scientist-astronaut by NASA in June 1965. Dr. Kerwin served as science-pilot for the *Skylab* 2 (SL-2) mission, which launched on May 25 and terminated on June 22, 1973. SL-2 was for the initial activation and 28-day flight qualification operations of the *Skylab* orbital workshop. Dr. Kerwin was the first physician selected to be an astronaut and the first US physician to go into space.

5. Roll up the tortillas and place, seam side down, in a 2-quart baking dish.
6. Pour remaining sauce over tortillas. Sprinkle remaining cheese on top.
7. Bake at 350°F for 20–30 minutes or until bubbly.

Yield: Serves 4–6

GERALD CARR'S CROCK POT CHILI

1 lb pinto, red, or anasazi beans
2 tbsp olive or vegetable oil
4 medium green peppers, chopped
3 medium onions, chopped
4 or more garlic cloves, minced
   (more is better!)
1 lb ground turkey
1 lb ground turkey sausage
1 package Wick Fowler's 2-Alarm
   Chili Mix™
OR

1/4 cup chili powder
1 tsp cayenne pepper
1 tbsp cumin
1 tbsp dried oregano
4 cups canned chopped or diced
   tomatoes
1 tbsp vinegar

1. Wash beans and soak overnight in water.
2. Sauté peppers in vegetable or olive oil; add onion, and cook until tender, stirring frequently. Add minced garlic.
3. Brown the meat, and stir in the spices. Then add the onion/pepper mixture and cook for about 10 minutes, stirring frequently.
4. Pour the meat mixture and the soaked beans into a crock pot. Add canned tomatoes and vinegar. Add water as desired, and salt to taste. Simmer for 18–24 hours. If you don't have a crock pot, do it in the oven at 250°F.

Meet the Astronaut: **Gerald Carr, Skylab 4 (84 days)**

Colonel Gerald Carr received a bachelor's degree in Mechanical Engineering from the University of Southern California in 1954, a Bachelor of Science degree in Aeronautical Engineering from the US Naval Postgraduate School in 1961, and a Master of Science degree in Aeronautical Engineering from Princeton University in 1962. He received flight training at Pensacola, Florida, and Kingsville, Texas, and was then assigned to Marine All-Weather-Fighter-Squadron 114, where he gained experience in the F-9 and the F-6A Skyray. After postgraduate training, he served with Marine All-Weather-Fighter-Squadron 122 from 1962 to 1965, piloting the F-8 Crusader in the United States and the Far East. Colonel Carr was one of 19 astronauts selected by NASA in April 1966. Colonel Carr was the commander of *Skylab 4* (third and final manned visit to the *Skylab* Orbital Workshop) launched November 16, 1973, and concluded February 8, 1974. This was the longest manned flight (84 days, 1 hour, 15 minutes) in the history of US manned space exploration at that time.

LINDA AND DICK GORDON'S CRAWFISH ETOUFFEE

1 stick (8 tbsp) unsalted butter
2 tbsp all-purpose flour
1 cup yellow onions, chopped
1/2 cup celery, chopped
1/2 cup green bell peppers, chopped
1/4 cup green onions, chopped
1 tbsp garlic, minced
2 bay leaves
1 tsp salt
1/4 tsp cayenne
2 tbsp dry sherry
1½ cups shrimp stock or water. Shrimp stock is not readily available but easy to make. Add shrimp peels and/or heads to a pot. Add chopped celery and onion if desired. Add water to cover and bring to boil. Simmer for 1 hour. Cool and strain. It will keep frozen for 3 months.
1 lb crawfish tails
2 tsp fresh lemon juice
3 tbsp chopped fresh parsley leaves, plus more for garnish
Cooked long grain white rice, as accompaniment

1. In a large pot, melt the butter over medium-high heat. Add the flour and cook, stirring, to make a light roux.
2. Add the onions, celery, bell peppers, green onions, garlic, bay leaves, salt, and pepper and cook, stirring, until the vegetables are soft, about 5 minutes.
3. Add the sherry and cook for 2–3 minutes. Add the stock and crawfish tails and bring to a boil.
4. Reduce the heat and simmer until thickened, about 5 minutes.
5. Add the lemon juice.
6. Stir in the parsley and remove from the heat.
7. Adjust the seasoning to taste. Serve over rice, garnished with additional parsley.

GERALD CARR'S TURKEY SAUTE

1. Ask your butcher to cut you some turkey cutlets from a breast of turkey. They should be cut parallel to the breast bone and about ¼ in. thick. That should give you slabs of meat about 2–3 in. wide by 4–6 in. long.
2. At home, pound the cutlets fairly gently until they are about 1/8-in. thick. Sauté them in extra virgin olive oil until they are golden brown. The olive oil should be laced with minced garlic, coarse ground pepper, and some lemon juice. Save the garlic pepper oil.
3. Serve the turkey with your favorite pasta and a green vegetable. Pour the garlic pepper oil over it all. Some favorite vegetables are asparagus and spinach, either grilled or steamed.

Meet the Astronaut: **Dick Gordon, Gemini XI and Apollo 12**

Captain Dick Gordon received a Bachelor of Science degree in Chemistry from the University of Washington in 1951. He received his wings as a naval aviator in 1953. He then attended All-Weather Flight School and jet transitional training and was subsequently assigned to an all-weather fighter squadron at the Naval Air Station at Jacksonville, Florida. In 1957, he attended the Navy's Test Pilot School at Patuxent River, Maryland, and served as a flight test pilot until 1960. Captain Gordon was one of three groups of astronauts named by NASA in October 1963. Captain Gordon has completed two spaceflights, logging a total of 315 hours and 53 minutes in space—2 hours and 44 minutes of which were spent in EVA.

PANDORA AND BOB CRIPPEN'S COWBOY BEANS

1 lb lean ground beef
1 medium yellow onion, finely diced
1 clove garlic, minced
1 16-oz can baked beans
1 16-oz can pinto beans, drained
  and rinsed

1 16-oz can light red kidney beans,
  drained and rinsed
1 16-oz bottle of your favorite
  barbecue sauce
1 cup jalapenos, finely chopped
  (optional)

1. Brown ground beef and drain.
2. Add onion, garlic, beans, barbecue sauce, and jalapenos, if
    desired.
3. Simmer 15 minutes. Serve with cornbread.

Yield: 16 servings

Meet the Astronaut: **Bob Crippen, STS-1, STS-7, STS-41C, STS-41G**

Captain Crippen received a Bachelor of Science degree in Aerospace Engineering from the University of Texas in 1960. He received his commission through the Navy's Aviation Officer Program at Pensacola, Florida. He received his wings at Chase Field in Beeville, Texas. In October 1966 he was selected for the USAF Manned Orbiting Laboratory Program and was later assigned to NASA when the MOL program was discontinued. Captain Crippen became a NASA astronaut in September 1969. He served as pilot on STS-1 (April 12–14, 1981) and was the spacecraft commander on STS-7 (June 18–24, 1983), STS-41C (April 6–13, 1984) and STS-41G (October 5–13, 1984). A four flight veteran, Crippen has logged over 565 hours in space. STS-1, the first shuttle mission, was also the first manned vehicle to be flown into orbit without benefit of previous unmanned "orbital" testing. It was also the first to launch with wings using solid rocket boosters and the first winged reentry vehicle to return to a conventional runway landing, weighing more than 99 t as it was braked to a stop on the dry lakebed at Edwards Air Force Base, California.

KEN REIGHTLER'S TILGHMAN ISLAND CRAB CAKES

1 egg, beaten
1 heaping tbsp mayonnaise
1 heaping tbsp bottled mustard
1 cup dried bread crumbs
1 tsp Worcestershire sauce

Juice of half a lemon
1 lb crab meat, picked over for
    shells
Butter or vegetable oil for cooking

1. Mix all ingredients except crab meat together in a bowl. Beat until well blended.
2. Place crab meat into a bowl; pour blended mixture over and mix gently but thoroughly.
3. Form into small cakes with hands, pressing together to prevent breakup during cooking.
4. The cakes may be broiled, brushing with butter while broiling, or pan fried in a skillet with butter or oil. Fry on both sides until browned.

Meet the Astronaut: **Ken Reightler, STS-48, STS-60**

Ken Reightler received a Bachelor of Science degree in Aerospace Engineering from US Naval Academy in 1973, and Master of Science degrees, in 1984, in Aeronautical Engineering from the US Naval Postgraduate School and in systems management from the University of Southern California. He graduated from the US Naval Academy in 1973 and was designated a naval aviator in August 1974 at Corpus Christi, Texas. He attended the US Naval Test Pilot School at Patuxent River, Maryland, where he graduated in 1978. He was serving as the chief flight instructor at the US Naval Test Pilot School when he was selected for the astronaut program in 1987. Ken Reightler was the pilot on STS-48 (1991) and STS-60 (1994) and has logged over 327 hours in space.

PAUL WEITZ'S WHITE CHILI

1 tbsp salad oil
1 medium-size onion, chopped
1 garlic clove, minced
1 tsp ground cumin
2 whole large chicken breasts,
    skinned, boned, and cut into
    bite-sized chunks
1 16- to 19-oz can of white kidney
    beans (cannelloni), drained
1 15½- to 19-oz can of garbanzo
    beans, drained

1 12-oz can white corn, drained
2 4-oz cans chopped mild green
    chilies
2 chicken-flavored bouillon cubes
    or 2 packets of dried bouillon
1½ cups water
Hot pepper sauce, to taste
Parsley sprigs for garnish
1 cup (1/4 lb) Monterey Jack
    cheese, shredded

1. Preheat oven to 350°F.
2. Heat salad oil in small saucepan over medium heat. Cook onion, garlic, and cumin until onion is tender.
3. In a 2 1/2-quart casserole dish, combine onion mixture with chicken, white kidney beans, garbanzo beans, corn, green chilies, bouillon, and water. Cover casserole and bake 50–60 minutes until chicken is tender.
4. To serve, stir hot pepper sauce into chili to taste. Garnish with parsley. Serve with shredded cheese.

Yield: 8 servings

Meet the Astronaut: **Paul Weitz, Skylab 2 and STS-6**

Paul Weitz received a Bachelor of Science degree in Aeronautical Engineering from Pennsylvania State University in 1954 and a Master's degree in Aeronautical Engineering from the US Naval Postgraduate School in Monterey, California, in 1964. He was awarded his wings in September 1956 and served in various naval squadrons until he was selected as an astronaut in 1966. He served as pilot on the crew of *Skylab-2* (SL-2), which was launched on May 25 and ended on June 22, 1973. SL-2 was the first manned Skylab mission and activated a 28-day flight. In logging 672 hours and 49 minutes aboard the orbital workshop, the crew established what was then a new world record for a single mission. Weitz logged 2 hours and 11 minutes in extravehicular activities. Weitz was also spacecraft commander on the crew of STS-6, which launched from Kennedy Space Center, Florida, on April 4, 1983. With the completion of this flight, Paul Weitz logged a total of 793 hours in space.

What You'll Find at Your Supermarket

Garlic herb Italian style baked tofu by White Wave Inc. NASA processes the tofu in a retort package before use.
Beef Ravioli in a retort pouch by the Wornick Company
Noodles and Stroganoff Sauce with Beef by Oregon Freeze Dry
Backpackers pantry freeze dried chicken cashew curry by American Outdoor Products
Cheese Tortellini in tomato sauce by SOPAKCO
Chicken breast strips with rib meat, chopped and formed with chunky salsa in a retort pouch by Wornick
Chicken with Black Beans in a retort pouch by SOPAKCO
Freeze Dried Leonardo da Fettuccine by Alpine Aire LLC
Seasoned chicken breast fillet in a retort pouch by SOPACKO

Individually quick frozen fully cooked roasted boneless pork loin by Rose Packing. NASA processes the pork chop in a retort package before use.

Stouffers frozen Macaroni and Cheese. NASA cooks and freeze dries before use

Mountain House freeze dried noodles and chicken by Oregon Freeze Dry

Rice and Chicken-Freeze dried rice and chicken by Oregon Freeze Dry

Premium skinless and boneless pink salmon by Chicken of the Sea International

Mountain House freeze dried seafood chowder by Oregon Freeze Dry

Freeze dried spaghetti with meat sauce by Oregon Freeze Dry

Mountain House freeze dried Oriental style rice and chicken with vegetables by Oregon Freeze Dry

Fancy albacore solid white tuna in spring water by Starkist Tuna Co

Premium chunk light tuna in spring water in pouches by Starkist

Tuna Salad Spread-Ready to eat tuna salad spread in cans by Bumble Bee

Mountain House freeze dried turkey tetrazzini by Oregon Freeze Dry

CHAPTER 7

Eat Your Vegetables!

It's time to say more about that "Big Question" mentioned in the introduction. Lavatory facilities were absent in the *Mercury*, *Gemini*, and *Apollo* capsules. The first lavatory in an American spacecraft did not appear until the *Skylab* space station was launched in 1973, twenty-two years after Alan B. Shepherd rode the first *Mercury* capsule into space. Shepherd would have appreciated the lavatory. No provisions for going to the bathroom were made for his flight. The idea was to stuff him into the capsule, light the engines, and pluck him out of the ocean, all in less than an hour. It didn't work out that way. Technical problems stretched Shepherd's time on the launch pad to hours, and the coffee he had before suiting up didn't seem like such a good idea to him after all. Ultimately, Shepherd, the first American astronaut to fly into space, did so in wet pants.

From that flight on until the early space shuttle program, every space suit featured a built-in UCD, or urine collection device. The UCD was a bag and a tube. Eventually, space suit designers concluded that adult-size diapers were much more convenient and more reliable to use. They are now a regular part of spaceflight.

In the very early missions, space dietitians were called on to minimize fiber content in food to reduce the need for bowel movements. Urinating into a bag was one thing, but bowel

C.T. Bourland, G.L. Vogt, *The Astronaut's Cookbook*, DOI 10.1007/978-1-4419-0624-3_7,
© Springer Science+Business Media, LLC 2010

movements were a whole level of magnitude more difficult. Forget dignity. Inside the *Gemini* and *Apollo* capsules, privacy was non-existent. Bowel movements involved pressing a plastic bag with adhesive strips . . . you get the idea. It was best if you did not have to go at all.

The *Apollo* food and beverage list did not contain any vegetables as entrées. There were a couple of meat and vegetable combinations on the list. With a lavatory installed on the *Skylab* space station, diets could be changed. Efforts were made to include vegetables, but freeze-drying was about the only preservation method available that would meet the weight and volume restrictions of the spacecraft.

Freeze-drying vegetables met with mixed results. Although the flavor and nutrients of vegetables were maintained, the texture was often compromised. They freeze dried beautifully but ended up with a very fragile structure that collapsed when handled. Some vegetables also presented color problems when freeze-dried. Freeze dried carrots turned white after a short storage period, and green beans were no longer green if exposed to light for lengthy periods. On the other hand, asparagus, green beans (if you don't mind a color change), peas, and corn freeze dried well. These foods have become staples in the space food systems. Freeze-dried spinach, broccoli, and cauliflower are also featured on space shuttle and ISS menus. Carrots and tomatoes (really a fruit) are popular space foods, but they must be thermostabilized to be acceptable.

Regardless of the skill in freeze-drying vegetables and packaging them with spices and other flavorings, the optimum preservation method for vegetables is ordinary freezing. *Skylab* had a freezer. In the early designs for the International Space Station, a habitation module was planned for one of the station's living areas. The module was later canceled, which was unfortunate for the space food laboratory people and the astronauts themselves. The habitation module would have included a food freezer. The possibilities for frozen food would have been endless (Figure 7.1). Perhaps freezers will be added to future space stations and lunar bases.

For now, vegetables are either freeze-dried or added to main dishes in a thermostabilized meat entrée. The Russian space program also has meat and vegetable combinations included on their food lists. Their canned products include vegetables with chicken, beef, and pork.

FIGURE 7.1 Sample meal from proposed ISS frozen food system (NASA photograph).

70 PERCENT

In spite of the personal preference menus, hot and cold water, and a reliable oven for heating food, the actual intake of food on shuttle missions average around 70 percent of the caloric requirement determined by dietitians and doctors. Although unappetizing food choices in the early program resulted in astronauts under-eating, other factors are at work on shuttle missions. Due to the short duration of shuttle missions and heavy workloads, crew members often do not have sufficient time to eat meals. Further, space adaptation syndrome or space motion sickness reduces food consumption in the first few days of shuttle flights. About half of all astronauts spend one or more shaky days adapting to microgravity.

CHARLES BOURLAND'S DIARY: TASTING THINGS OF UNKNOWN ORIGIN

One of the most pleasant—and unpleasant—experiences of being involved in the development of space food is the opportunity (or requirement) to taste food samples. Particularly in the early days food companies took note of the TV advertisements for Tang™ and Space Sticks and wanted their food to go into space, too. Sometimes they called ahead, other times they just mailed it in or personally brought it in. It is unbelievable how many kinds of beef jerky and BBQ sauce are on the market. Many are made in a garage, and all chefs believe theirs to be the best. Many of the samples were good, but just not adaptable to spaceflight. I always felt obligated to at least evaluate the sample and give a report. If the sample was from an unknown source, we usually sent it to the microbiology laboratory before tasting. We received several "nutritious" algae samples, but I have never tasted an algae sample that was not bitter and ghastly.

Once, a restaurant in Philadelphia arranged to send a Hoagie sandwich to us for evaluation. A two-day-old sandwich arrived in the middle of summer in Houston. There wasn't any attempt to keep it cold. I called the chef and told him I was afraid to taste it because of the lack of refrigeration. It had all the components for a good case of food poisoning. He said not to worry, he sent them out all the time and no one had gotten sick or died. I still refused to taste it and disposed of it. Later, he wrote to his congressman complaining that I refused to taste his sandwich. NASA had to respond to a congressional inquiry—lots of paperwork!

THE WCS1

For some reason, NASA never likes to use direct language when a technical-sounding acronym can be used. WCS means waste collection system—the toilet. Space toilets operate on a vacuum principle. Urine is drawn into a tube with air suction. Solid waste is also drawn away with air suction, and it ends up in a tank where it is dried and compacted. Although looking more like a Rube Goldberg device than a traditional toilet, the WCS does have a seat and a urine tube that resembles part of a goose-neck lamp. There are levers and switches and spring-mounted thigh bars that keep crew members from drifting away from the seat (Figure 7.2).

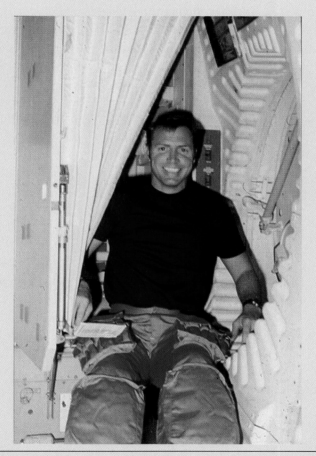

FIGURE 7.2 Astronaut David Walker demonstrating the Shuttle potty. It works by using air instead of water to move materials (NASA photograph).

PEA BARS

NASA and the US Army's Natick Soldier System Center have always worked closely. They share many of the same food development goals, such as extended shelf life and consumption in extreme situations. It has been a beneficial, although not always successful, collaboration. Natick developed some new vegetable bars. NASA immediately saw the potential and flew them on the American/Soviet *Apollo-Soyuz* mission in 1975. The bars were made by compressing partially dried spinach or peas under high pressure. They were then freeze-dried to remove any remaining moisture. In the end the spinach and peas retained their bright color and looked delicious. Their small size made them ideal for stowing in the cramped *Apollo* capsule. All the crew needed to do was to rehydrate. Unfortunately, rehydration is a lot harder to do well in microgravity than on Earth. The taste and texture was "less than desired," and pea and spinach bars never flew again.

SS ASPARAGUS (FIGURE 7.3)

2 lb fresh asparagus   3¼ tsp salt
4 cups water   1/4 tsp monosodium glutamate

1. Rinse asparagus in running water and cut/break into 1- to 2-in. pieces.
2. Add salt and monosodium glutamate to water and bring to boil.
3. Add asparagus and cook for 11–15 minutes, until tender.
4. Drain and serve.

Yield: 6 servings

Note: NASA further processes the Asparagus by freeze drying prior to use in space.

If you were making the asparagus for space food, you would have to break the asparagus by hand rather than cutting it up with a knife. The first shipment of *Skylab* asparagus from the contractor was cut,

FIGURE 7.3 Asparagus.

and it turned out to be tough and stringy. The shipment was rejected, and the contractor had to produce more that was acceptable.

SS BLACK BEANS (FIGURE 7.4)

29 oz canned black beans, liquid reserved
4 oz canned crushed tomatoes
1/2 tsp ground cumin
1/4 tsp oregano
1/4 tsp black pepper

1/4 tsp salt
3 tbsp canned green chilies, roasted, peeled, and diced
1 medium onion, diced
1/2 tsp garlic, chopped

1. Drain and measure liquid from beans; save half of the liquid for later use.
2. Combine crushed tomatoes, including liquid, cumin, oregano, black pepper, and salt and mix well.
3. Combine tomato mixture, green chilies, including liquid, onions, garlic, drained black beans, and reserved bean juice.
4. Heat to a boil and cook until onions are tender. Serve warm.

Yield: 6 servings

Note: NASA further processes the Black Beans by thermo-processing in a retort pouch.

FIGURE 7.4 Black beans.

SS CARROT COINS

1/3 cup water
1 tsp National 150 filling aid starch
    from National Starch and
    Chemical Co (You may substitute
    cornstarch for this)

1/4 tsp salt
1 tbsp unsalted butter
1½ lb of carrots, sliced into rounds

1. Combine starch and salt with 1 tbsp of the water.
2. Add remaining water and butter to a saucepan and heat to melt butter.
3. Add starch and salt mixture, and mix well.
4. Add carrots, mix well, and heat on medium high, covered, until carrots are tender, about 15 minutes.

Yield: 6 servings

Note: NASA further processes the Carrot Coins by thermo-processing them in a retort pouch.

SS CORN (FIGURE 7.5)

1 lb package frozen corn
3 tbsp butter flavor granules such
   as Butter Buds Sprinkles™

2½ tsp instant CLEARJEL starch
   from National Starch and
   Chemical*
1 tsp dried parsley

Cook corn per manufacturer directions. Add butter granules, starch, and parsley and mix well and serve.

Yield: 5 servings

*The added starch aids consumption in microgravity and may not be needed for 1 G consumption.

Note: NASA produces SS Corn by freeze drying the corn and then adding each dry ingredient to the individual serving package

FIGURE 7.5 Corn.

SS GREEN BEANS AND POTATOES

1½ tsp National 150 filling aid starch from National Starch and Chemical Co. Can substitute cornstarch
1 tsp salt
1/4 tsp coarse ground black pepper

1/2 cup water
2 tbsp unsalted butter
1 16 oz package frozen green beans
1 cup diced red potatoes

1. Mix starch, salt, and pepper with 2 tbsp of the water and mix well.
2. Put remaining water and butter in a saucepan and heat to melt butter.
3. Add green beans and potatoes and heat to boiling; simmer until beans and potatoes are fork tender.
4. Add starch, salt, and pepper mixture to beans, mix well, and simmer for 2 minutes.

Yield: 6 servings

Note: NASA further processes the Green Beans and Potatoes by thermo-processing in a retort pouch.

SS HOMESTYLE POTATOES

3 cups diced hash-brown potatoes
1/3 cup red bell peppers, diced
1/4 cup green bell peppers, diced
2 tbsp fire-roasted Anaheim green chili peppers, diced

1/3 cup onions, diced
2 tbsp olive oil
3/4 tsp seasoned salt
2 tsp ground black pepper

1. Combine potatoes, red and green bell peppers, green chilies, and onions in a saucepan and mix well.
2. Combine oil, salt, and pepper, mix well, and add to the potato mixture.
3. Toss and coat potatoes evenly. Heat on medium heat, stirring occasionally until potatoes are tender.

Yield: 6 servings

Note: NASA further processes the Homestyle Potatoes by thermo-processing in a retort pouch.

SS MIXED VEGETABLES

1½ lb frozen mixed vegetables
1/2 tbsp National 150 filling starch
   from National Starch and
   Chemical Co. Not required
   unless thermoprocessing.

1/2 tsp salt
1/4 tsp coarse ground black pepper
1/3 cup water
1 tbsp unsalted butter

1. Thaw mixed vegetables in refrigerator in advance.
2. Combine starch, salt, and pepper with 1 tbsp water. Add remaining water and butter to a saucepan and heat to melt butter.
3. Add mixed vegetables, mix well and heat on medium heat with cover until vegetables are tender. Add starch, salt and pepper mixture and mix well, and simmer for 2 minutes.

Yield: 6 servings

Note: NASA further processes the Mixed Vegetables by thermo-processing in a retort pouch.

SS POTATO MEDLEY (FIGURE 7.6)

2 tbsp olive oil
2 tbsp balsamic vinegar
1/2 tsp fresh thyme, chopped
2 tsp garlic, chopped
2 oz fat-free chicken
   broth

2 cups 1-in. sweet potatoes, peeled
   and cubed
2 cups 3/4-in. russet potatoes,
   peeled and diced
1½ cups skin-on 3/4-in. red
   potatoes, diced

1. Preheat oven to 450°F.
2. Combine olive oil, balsamic vinegar, thyme, garlic, and chicken broth; mix well.
3. Add oil mixture to potatoes, and toss to coat. Place potato mixture in a baking dish and bake until potatoes are tender, approximately 30 minutes. Serve hot.

Yield: 6 servings

FIGURE 7.6 Potato Medley.

Note: NASA further processes the Potato Medley by thermo-processing it in a flexible pouch.

SS RICE PILAF

4 cups water
2 cups Riviana Foods Foodservice
   Rice Pilaf

3 tbsp butter

Cook rice pilaf, butter, and water per manufacturer's directions and serve.

Yield: 8 servings

Note: NASA further processes the Rice Pilaf by freeze drying before use.

SS SPICY GREEN BEANS

2 tsp vegetable oil
2 tbsp fresh jalapenos, diced
5 tsp garlic, chopped
2 tsp ground turmeric
2 tsp ground cumin
1/2 tsp cayenne pepper
1½ lb frozen extra fine whole green
  beans (haricots verts). May
  substitute with fresh.

2 tsp salt
2½ tbsp water
2 tbsp sesame seeds
1 tsp Ultra-Sperse M modified
  starch from National Starch and
  Chemical Co. Not required for
  home use.

1. Heat saucepan and add vegetable oil.
2. Add chopped jalapeno and garlic to saucepan and sauté.
3. Add turmeric, cumin, and cayenne and mix vigorously.
4. Add green beans, salt, and water. Mix well to coat the green beans with the spice mixture.
5. Heat to simmer, add sesame seeds, mix well, and serve.

Yield: 6 servings

Note: NASA further processes the Spicy Green Beans by freeze drying before use.

EMERIL'S SPICY GREEN BEANS WITH GARLIC (FIGURE 7.7)

1/4 cup clarified butter or vegetable oil
3 cloves garlic, thinly sliced
2 small green peppers (such as
  jalapeno or serrano), stems and
  seeds removed, minced
2 tsp turmeric powder

2 tsp ground cumin
1/8 tsp cayenne
1 lb green beans, tough ends
  removed
1/4 cup water
1½ tsp salt
3 tbsp sesame seeds

1. In a large sauté pan, heat the butter over medium-high heat.
2. Add the garlic slivers, peppers, turmeric, cumin, and cayenne, and cook, stirring, until the garlic begins to turn golden, about 2 minutes.
3. Add the green beans, water, salt, and stir well.

FIGURE 7.7   Emeril's spicy green beans with garlic.

4. Cover and cook over medium-low heat, stirring occasionally, until the beans are tender, 4–5 minutes.
5. Add the sesame seeds and cook, uncovered, stirring until toasted, 2–3 minutes.
6. Remove from the heat and adjust seasoning to taste.

Yield: 4 servings

Note: When used in space NASA further processes the Spicy Green Beans with Garlic by freeze drying.

SS SQUASH CASSEROLE

1½ lb frozen squash, sliced (may substitute fresh)
1 medium onion, diced
1 tbsp unsalted butter
2 tbsp egg substitute such as Egg Beaters™

1/2 tsp salt
1/4 tsp ground black pepper
1 tbsp sugar
1/4 cup packaged cornbread stuffing

1. Preheat convection oven to 500°F. Place squash into a baking dish that has been lightly sprayed with non-stick vegetable cooking spray.
2. Roast squash for approximately 30 minutes or until squash starts to brown.
3. Place onions into a baking dish that has been lightly sprayed with nonstick vegetable cooking spray.
4. Roast onions for approximately 10 minutes, or until onions start to brown.
5. Combine roasted onions with roasted squash and butter. Mix until the butter is melted and distributed evenly.
6. Combine the egg product, salt, black pepper, and sugar and set aside. Combine squash mixture, egg mixture, and cornbread stuffing.
7. Place in a baking dish and heat at 350°F for 10 minutes.

Yield: serves 6

Note: NASA further processes the Squash Casserole by retorting in a flexible pouch.

RACHAEL RAY'S VEGETABLE CURRY IN A HURRY

3 tbsp olive oil, divided
1½ cups Basmati rice
4 cups chicken or vegetable stock, divided
1 bay leaf, fresh or dried
Zest of 1 lemon
1 tsp turmeric
2 tsp coriander
2 tsp cumin
1/2 tsp cardamom, optional

1 tbsp butter
1 medium onion, thinly sliced
3 cloves garlic, chopped
1 small head cauliflower, chopped
1 firm eggplant, peeled and chopped (You can peel away only half of the skin if you like the color and texture.)
1 red bell pepper, seeded and chopped

1 14-oz can diced tomatoes, drained
1 15-oz can of chick peas, drained
Salt and pepper
3 tbsp mild or hot curry paste
3 tbsp mango chutney

3 scallions, chopped, for garnish
A handful of cilantro or parsley, for garnish
Toasted slivered almonds or pieces of cashew, for garnish

1. Heat a medium pot over medium heat with olive oil. Add rice and toast a minute or two.
2. Add 3 cups chicken or vegetable stock and the bay leaf, lemon zest, turmeric, coriander, cumin, and cardamom.
3. Cover pot and bring rice to a boil. Reduce the heat and simmer for 18 minutes.
4. Fluff rice with a fork, remove bay leaf, and add butter. Toss to coat the rice evenly, then serve.
5. While rice cooks, make the vegetables. Heat a deep non-stick skillet over medium high heat with 2 tbsp of olive oil.
6. Add onion, garlic, cauliflower, eggplant, and bell pepper. Cover and cook, stirring occasionally for 7–8 minutes, until tender.
7. Uncover and add the tomatoes, chick peas, salt, pepper, curry paste, chutney, and remaining cup of chicken or vegetable stock. Simmer 6–7 minutes longer.

Serve curry with scoops of rice on top. You can use an ice cream scoop to portion it. If you put the rice on top, it will not get mushy. Garnish with scallions, cilantro, or parsley and nuts.

Yield: 4 servings

Note: When used in space NASA further processes the Vegetable Curry by freeze drying.

SS SUGAR SNAP PEAS

3½ cups fresh stringless sugar snap peas
1/3 cup water
1 tbsp butter
2/3 tsp salt

1 tsp National 150 filling starch (or cornstarch)
1/4 tsp ground black pepper

1. Add all ingredients except the starch to a medium pot.
2. Heat to boiling, reduce heat to simmer, and cook until peas are tender.

3. The starch is used for retorting and not essential; however, if desired, add a teaspoon of cornstarch to 3 tbsp of water and add to the heating mixture.

Yield: 6 servings

Note: NASA further processes the Sugar Snap Peas by thermo-processing in a flexible pouch.

SS TOFU WITH HOT MUSTARD SAUCE

1 Napa cabbage
18 oz tofu extra firm silken style
1/3 cup balsamic vinegar
1/4 cup plus 1 tbsp hot bottled mustard
1/2 tsp peanut oil
dash cayenne pepper

2 tsp dried onions, minced
4 tsp garlic, minced
1 tbsp modified food starch (cornstarch may be substituted)
1/4 cup unsulfured molasses

1. Wash cabbage thoroughly. Trim cabbage ends and cut into 1-in. by 2-in. pieces. Allow cabbage to dry.
2. Preheat oven to 350°F.
3. Cut tofu into1/2-in. cubes. Transfer tofu to a baking sheet that has been sprayed with a vegetable cooking spray.
4. Bake tofu for 12–15 minutes and set aside.
5. Add ¼ cup balsamic vinegar, hot mustard, peanut oil, cayenne pepper, onion, and garlic to a saucepan and mix well.
6. Add cabbage and heat to simmer, stirring occasionally.
7. Combine starch and remaining balsamic vinegar, stir well, and add to the saucepan. Simmer for 3–5 minutes.
8. Add molasses to the mixture and mix well. Add tofu and mix gently.
9. Continue to heat until cabbage is tender, about 15 minutes.

Yield: 6 servings

Note: NASA further processes the Tofu with Hot Mustard Sauce by retorting in a flexible pouch.

SS TOFU WITH HOISIN SAUCE

1 Napa cabbage
18 oz tofu extra firm silken style
1 tbsp low sodium soy sauce
1 tbsp modified food starch
   (cornstarch may be substituted)
1/4 cup rice wine

1 tbsp ketchup
3 tsp garlic, minced
4 tsp light brown sugar
1/3 cup Hoisin sauce (available in
   the Asian section of your
   supermarket)

1. Wash cabbage thoroughly. Trim ends and cut into 1-in. by 2-in. pieces. Allow cabbage to dry.
2. Preheat oven to 350°F.
3. Cut tofu into 1/2-in. cubes. Transfer tofu to a baking sheet that has been sprayed with a vegetable cooking spray.
4. Bake tofu at 350°F for 12–15 minutes and set aside.
5. Combine the soy sauce and starch and set aside. Add rice wine to a saucepan followed by the ketchup and garlic.
6. Heat to simmer, stirring occasionally.
7. Stir the starch slurry and add to the saucepan.
8. Continuously mix and simmer for 3–5 minutes.
9. Add brown sugar and Hoisin sauce and mix to combine.
10. Add tofu and mix gently.
11. Add cabbage, cover, and heat until cabbage is tender, about 15 minutes.

Yield: 6 servings

Note: NASA further processes the Tofu with Hoisin Sauce by retorting in a flexible pouch.

SS TOMATOES AND EGGPLANT

1½ tbsp water
1 tbsp modified food starch or
   cornstarch
1 16.5-oz can of diced tomatoes
1 cup diced zucchini
1 medium yellow onion, diced
3/4 cup peeled and diced eggplant
1/3 cup diced red pepper
1/3 cup diced green pepper
1/4 cup tomato juice

2 tbsp tomato paste
2 tbsp olive oil
2 tsp chopped garlic in water (may
   substitute fresh)
1 ½ tsp salt
1/2 tsp coarse ground black pepper
1/4 tsp dried basil
1/4 tsp dried oregano
Pinch of coriander, ground
   coriander seed

1. In a container mix water and starch and set aside.
2. In a saucepan combine all other ingredients and mix well.
3. Add starch mixture and bring to boil. Cook 5–7 minutes, until eggplant and zucchini are tender.

Yield: 8 servings

Note: NASA further processes the Tomatoes and Eggplant by placing it in a retort package and thermally processing. Overcooking is required to make this recipe similar to the NASA product.

The following are some of the astronauts' own favorite recipes.

KEN REIGHTLER'S CORN PUDDING

1 16-oz can of corn, cream style
1 16-oz can of corn, whole kernel style
3 eggs, beaten
1 13-oz can of evaporated milk

1 tbsp sugar
1 tsp salt
2 tsp cornstarch or flour
Garlic and lemon to suit your taste
2 tbsp butter

Mix ingredients together and dot butter on top. Bake at 350°F in a 2-quart casserole for about 2 hours (or until a knife can be inserted and comes out clean). Serve warm. Great reheated!

RHEA SEDDON'S DILLED GREEN BEANS

2 cups sugar
1 cup apple cider vinegar
4 tbsp vegetable oil

2 tbsp onion, finely chopped
1 tbsp mixed dill seed and dill weed
4 cans whole green beans

1. Combine sugar, vinegar, vegetable oil, onion, and mixed dill seed and weed in a saucepan and boil for 5 minutes.
2. Pour over the green beans and marinate in the refrigerator for at least 3 days, stirring occasionally. Serve cold or at room temperature.

Simple, quick, delicious—just what a busy astronaut mom needs!

Meet the Astronaut: Rhea Seddon, STS-51D, STS-40, STS-58

Dr. Rhea Seddon received a Bachelor of Arts degree in physiology from the University of California, Berkeley, in 1970, a doctorate of medicine from the University of Tennessee College of Medicine in 1973, and was selected as an astronaut candidate in 1978. She was among the first females selected for the astronaut program. A three-flight veteran with over 722 hours in space, Dr. Seddon was a Mission Specialist on STS-51D (1985) and STS-40 (1991), and was the Payload Commander on STS-58 (1993). Dr. Seddon has an interest in nutrition and was a member of the design and development team for the shuttle flight food hardware. She is married to former astronaut Robert L. Gibson.

MILLIE HUGHES-FULFORD'S EGGPLANT

1 large eggplant, washed and cut into disks
Cooking spray

Garlic salt

1. Spray one side of the eggplant slices with cooking spray.
2. Shake a little garlic salt (such as Lowery's with parsley) on sprayed side.
3. Place sprayed side down on stovetop grill plate (or large skillet) on medium high heat. While cooking lightly spray cooking spray on unsalted side of eggplant just before turning over (about 2 minutes, or until cooked)
4. Grill other side about 2 minutes.
5. Serve.

Meet the Astronaut: Millie Hughes-Fulford, STS-40

Dr. Hughes-Fulford entered college at the age of 16 and earned her BS degree in Chemistry and Biology from Tarleton State University in 1968. She completed her doctorate degree at Texas Woman's University in 1972. She was selected as a Payload Specialist by NASA in January 1983. Dr. Hughes-Fulford flew in June 1991 aboard STS-40 *Spacelab* Life Sciences (SLS 1), the first *Spacelab* mission dedicated to biomedical studies. The SLS-1 mission flew over 3.2 million miles in 146 orbits, and its crew completed over eighteen experiments during a nine-day period, bringing back more medical data than any previous NASA flight.

What You'll Find at Your Supermarket

Frozen broccoli and cheese by Birds Eye Foods. NASA cooks and freeze dries before use.

Parboiled long grain brown rice in retort pouches by Lambert Street Packaging.

Frozen cauliflower with cheese sauce by Green Giant plus frozen cauliflower florets. NASA cooks and freeze-dries before use.

Green Giant frozen creamed spinach by General Mills. NASA cooks and freeze-dries before use.

Frozen Garden Blend Italian Style vegetables by Kroger. Balsamic vinaigrette dressing by Kraft. NASA freeze-dries before use.

Dehydrated potato pearls by Basic American Foods.

Stouffers frozen potatoes au gratin by Nestle. NASA cooks and freeze-dries before use.

Rice with Butter MRE by Wornick Foods.

Create a meal stir fry teriyaki by Green Giant. NASA cooks and freeze dries before use.

Freeze dried brown rice with vegetables and mushrooms by Adventure Foods.

CHAPTER 8

Desserts

What would travel in space be like without dessert? Certainly less appealing. Most astronauts have the same cravings for sweets as the rest of us. The formula is simple. Astronauts plus desserts equal happy crews. OK, maybe that is a bit of an exaggeration. Spaceflight itself is enough to make any astronaut happy, but desserts do contribute to missions by helping crew unwind after intensive days. The problem is that not all desserts are equal. M&Ms might be the perfect complement to a meal one evening, but the next day it has to be cake!

Some desserts pose real challenges for space—especially cake. Though light in weight, a real plus for spaceflight, cake's volume is a problem. Cake is full of bubbles. If you flatten a slice of cake in a vacuum bag many of its charms will be destroyed. NASA's food laboratory staff spent a great deal of time experimenting with cakes and was able to extend their shelf life by irradiation. Of course, purely for scientific purposes, they had to sample them frequently to ensure their flavor. The one thing they could not overcome was the packaging and stowage problems.

Flight crews are not totally cake deprived. During their brief stay in the quarantine quarters at the Kennedy Space Center, a beautiful cake, decorated with their crew patch, is prepared by the quarantine facility staff. Some crews gobble it up on the spot, and others decide to eat it upon their return from space during a post-landing party.

C.T. Bourland, G.L. Vogt, *The Astronaut's Cookbook*, DOI 10.1007/978-1-4419-0624-3_8,
© Springer Science+Business Media, LLC 2010

By default, many of the space food desserts are commercially produced. Discovered in commercial vending machines, cellophane-wrapped desserts have excellent shelf life properties. Though not as tasty as fresh-baked cake, there is no such thing as a bad dessert in space. All that is necessary is to repackage them into flight approved packages.

One popular desert is the commercial individual serving of pudding. These, one would think, would not have to be repackaged. They should be the food laboratory's version of a slam-dunk. Not so. NASA flight and safety rules, by necessity, can be withering. You don't just say, "Fly this." Individual pudding servings demonstrate what can happen during the flight certification process.

Commercial puddings were first marketed in steel cans. The cans had to be measured, heated, probed, and otherwise abused to see what would happen to them in potential worst-case spaceflight scenarios. They even had to pose for engineering drawings before they could be approved. All space food items have special drawings made, just as if they were a bolt or other piece of flight hardware. After the steel can pudding servings were approved and flown, the companies making the puddings got the bright idea to package them in aluminum cans. That meant starting over and going back to the drawing board—literally. After recertification, the pudding companies got another bright idea—plastic. Back to the drawing board again!

Ice cream, though rare in space, is probably the most popular astronaut dessert when it is available. The *Skylab* space station had a freezer, and one of the frozen foods it held was ice cream. On occasion, small freezers have been flown on the space shuttle. Freezers are sometimes needed for medical and scientific experiment samples that have to be preserved exactly "as is" for analysis back on Earth. Purely to ensure that these freezers are ready to receive samples in space, crews insist they be powered up prior to launch. Of course, freezers are more energy efficient when they are filled. Ice cream is the ideal energy-saving freezer filler. Right!

CHARLES BOURLAND'S DIARY: WHAT GOES AROUND COMES AROUND

With the replacement of the plastic box with easily compressed vacuum sealed plastic bags for food items, the shuttle food tray design seemed like overkill. It was a bulky tray with friction slots for boxes, and it didn't stow well in the galley. A group of shuttle astronauts came to the food lab to discuss their concept for a new and improved food tray. The idea was that the tray would be a flat piece of metal with Velcro patches and long springs to hold the food, beverages, and condiments on the tray. I asked to be excused for a moment and went to the back room and pulled out a 1975 *Apollo-Soyuz* meal tray; flat metal, Velcro patches, springs and all, and showed it to them. After a few jaw drops, they responded "Yes, that is what we had in mind. Where did you get that?" It was a priceless moment. By the way, similar trays are still being used! (Figure 8.1).

FIGURE 8.1 Astronaut William MacArthur sipping a beverage with the current Shuttle meal tray strapped to his leg (NASA photograph).

SALT AND IRON

Astronauts tend to be a very healthy group. It can take years of training before being assigned to a mission. Maintaining health while waiting to fly is of great importance, and astronauts do their best to eat properly and get plenty of exercise. In space, astronauts eat nutritionally balanced menus and continue to exercise. There are some changes, however. Space menus tend to be higher in sodium and iron. This can be partially attributed to the use of many high sodium commercially product-based menu items and, in particular, commercial bread and cereal products enriched with iron. The requirement for shelf stable foods also affects sodium content, because sodium helps with preservation as well as enhancing flavor. But lower sodium intakes are desirable for astronauts in space to reduce bone loss. Less iron is needed due to normal reduced blood volume and red blood cell turnover that occurs in microgravity. So a balance must be achieved.

WANDERING M&MS

Wanting a bedtime snack, shuttle astronaut Bill Thornton quietly opened a package of M&M's. Some got away from him in the dimmed cabin light. In an M&Ms' version of billiards, the candies kept pinging him in the face as he was trying to sleep.

ASTRONAUT ICE CREAM

You will find silvery pouches of "Astronaut Ice Cream" over-flowing in bins in just about every science and aerospace museum in the country. For a couple of dollars, you, too, can enjoy the dessert sensation of the space program. Well, not quite. Inside the package is a dry and brittle brick of freeze-dried Neopolitan ice cream.

Freeze-dried ice cream is a real space food that never made it big in space. It was developed by the Whirlpool Corporation by special crew request for the *Apollo* program. Astronauts missed ice cream on long missions. Pouches of freeze-dried ice cream were flown on the *Apollo 7* mission in 1968. It was the first and last time it flew in space. The dry bricks softened with saliva and tasted remarkably like ice cream, but the creamy, icy sensation of regular ice cream was missing. It just wasn't popular with flight crews. That's when a marketing genius got the idea of selling astronaut ice cream to the public. The rest is history. It has become a top-selling item in museum gift shops all around the world.

LYOPHILIZATION, OR HOW TO MAKE ASTRONAUT ICE CREAM

Freeze-drying, or lyophilization, is a process that removes moisture from food. To make astronaut ice cream, start with a thick slice of your favorite ice cream. Place it in a vacuum chamber and start pumping out the air to form a partial vacuum. Then, turn on the heat to increase the vaporization rate. The water in the ice cream begins to vaporize and is trapped by a freezing coil. The process is slow, so have a good book handy. After several hours, all of the moisture in the ice cream has been extracted, and a crunchy brick of dry ice cream remains.

BLUE TONGUE SYNDROME

Astronauts have always had a shaky relationship with flight surgeons. Although no astronaut will ever disagree that flight surgeons are important to their well being, flight physicals are approached with some trepidation. Flight surgeons can ground astronauts if they suspect medical problems. *Mercury* astronaut Deke Slayton missed his *Mercury* flight due to a heart condition, and Alan Shepherd was grounded after his flight for an ear problem. Both waited many years before they were returned to flight status. Ken Mattingly was prevented from flying on *Apollo 13* (mixed blessing) because of his exposure to German measles. He never got the disease.

During their final pre-launch physicals, an entire space shuttle crew was told by a flight surgeon that they would be grounded. Something was going through the crew. They all had blue tongues. The surgeon was right about something going through the crew. Just prior to their physicals, they had gobbled up their mission cake (with blue frosting) at the quarantine quarters!

SS APRICOT COBBLER

9-in. frozen pie crust
18 oz frozen sliced apricots
1/4 cup water
1½ tsp natural strength lemon juice

3/4 cup plus 1 tbsp extra fine granulated sugar
2 tbsp quick cooking tapioca

1. Preheat oven to 400°F and bake piecrust approximately 12 minutes, or until golden brown.
2. Allow piecrust to cool, and break into ¼ in. pieces and set aside.
3. Add the apricots, water, and lemon juice to a saucepan and heat on medium while stirring.
4. After apricots are thawed, add the sugar and tapioca and heat to boiling while stirring.
5. Continue heating 12–15 minutes until apricots are tender.
6. Cool and add the piecrust pieces and mix well.

Yield: 6 servings

Note: NASA further processes the Apricot Cobbler by placing the mixture and crust in a retort pouch and thermo-processing.

SS BREAD PUDDING (FIGURE 8.2)

5½ oz whole loaf French bread
1¾ cup skim milk
3/4 cup egg substitute (such as Egg Beaters™)

1 tbsp pure vanilla extract
1 cup plus 1 tbsp extra fine granulated sugar
Pinch ground cinnamon
4 tbsp unsalted butter

1. Cut French bread into 1 in. cubes and set aside.
2. Add milk, eggs, and vanilla to the saucepan, stir, and begin heating on medium heat.
3. Add sugar and cinnamon and continue stirring.
4. Melt butter and add to the mixture.
5. Add the bread cubes and mix well.
6. Transfer to a baking dish coated with vegetable spray and bake at 325°F for 50–55 minutes, or until the top is light golden brown.

Yield: 6 servings

Note: NASA does not do the baking step, but further processes the Bread Pudding by placing the bread cubes and custard in a retort pouch and thermo-processing it in the pouch.

FIGURE 8.2 Bread pudding.

SS BUTTERSCOTCH PUDDING IN POUCHES

2 cups whole milk
1 cup water
3/4 cup extra fine granulated
   sugar
4 tbsp modified food starch. May be
   substituted for with cornstarch
1/2 tsp natural and artificial
   butterscotch flavor

1/4tsp salt
2 drops caramel artificial, liquid
   caramel color #525 from D.D.
   Williamson
1 drop yellow artificial liquid egg
   color

1. Combine starch, sugar, and salt. Mix well.
2. Add 1½ cup milk and water to a saucepan and heat to boiling.
   Remove from heat.
3. Add remaining milk to the starch mixture, mix well and add
   to the heated milk mixture.
4. Add yellow color, caramel color, and butterscotch flavor.
5. Mix well and transfer to serving dishes and chill.

Yield: 6 servings

Note: NASA further processes the Butterscotch Pudding by
thermo-processing in a retort pouch.

SS CHERRY BLUEBERRY COBBLER

1 cup yellow cake mix
3 tsp pecans, chopped
2 tbsp unsalted butter
1 tbsp modified food starch
   (cornstarch may be substituted)
2 tbsp lemon juice, natural
   strength

12 oz dark sweet cherries
7 tbsp extra fine granulated
   sugar
8 oz blueberries
1¼ tsp almond extract, pure or
   imitation

1. Preheat oven to 350°F.
2. Combine cake mix and chopped pecans in a bowl.
3. Melt butter and add to the cake mixture. Blend until mixture is moist and cohesive.
4. Press blended mixture evenly onto baking sheets lined with foil. Do not exceed ¼ in. depth.
5. Bake for 8 minutes, or until golden brown. Remove cooked dough with foil and allow to cool.
6. Cut cooled crust into approximately 2-in. by 3-in. pieces.
7. Combine starch and lemon juice to make a slurry.
8. Add cherries to a saucepan and heat until juice begins to release. Add sugar to cherries and mix well. Heat cherry-sugar mixture to simmer, stirring occasionally.
9. Stir the starch slurry and add to the cherry-sugar mixture.
10. Add blueberries and continue heating and stirring to simmer; hold for 3–5 minutes. Add almond extract.
11. To serve, pour the cherry mixture over the crust pieces.

Yield: 6 servings

Note: NASA further processes the Cherry-Blueberry Cobbler by placing the mixture and crust in a retort pouch and thermo-processing.

SS CHOCOLATE PUDDING CAKE (FIGURE 8.3)

3/4 cup Devil's Food cake mix
3 tbsp unsalted butter
1¾ cups skim milk
1 tsp vanilla extract
2 tbsp modified food starch
   (cornstarch may be substituted)

1/2 cup unsweetened cocoa
3/4 cup plus 2 tbsp extra fine
   granulated sugar
2 oz semisweet chocolate chips

1.   Preheat oven to 350°F.
2.   Transfer cake mix to a bowl.
3.   Melt 2 tbsp butter and add to the cake mixture.
4.   Blend until mixture is moist and cohesive.
5.   Press blended mixture evenly onto baking sheets lined with foil. Do not exceed ¼ in. depth.
6.   Bake for 8 minutes. Remove cooked dough with foil and allow to cool.
7.   Cut cooled crust into approximately 2-in. by 3-in. pieces.
8.   Combine ¼ cup milk, vanilla, and starch. Mix well and set aside.
9.   Add remaining milk, sugar, cocoa, and butter to a saucepan. Mix well and begin heating on medium heat.
10. Add chocolate chips to pan, stirring until melted.

FIGURE 8.3 Chocolate pudding cake.

11. Stir starch slurry and add to the mixture. Heat to boiling and remove from heat.
12. To serve, pour chocolate sauce over the crust pieces.

Yield: 6 servings

Note: NASA further processes the Chocolate Pudding Cake by placing the mixture and crust in a retort pouch and thermo-processing.

SS CHOCOLATE PUDDING IN POUCHES

3 tbsp modified food starch
   (cornstarch may be substituted)
3 tbsp cocoa powder
3/4 cup extra fine granulated sugar

¼ tsp salt
2 cups whole milk
1 cup water

1. Combine starch, cocoa, sugar, and salt. Mix well.
2. Add 1½ cups milk and water to a saucepan, heat to boiling, and remove from heat.
3. Add remaining milk to the starch mixture, mix well, and add to the heated milk mixture.
4. Mix well, transfer to serving dishes, and chill.

Yield: 6 servings

Note: NASA further processes the Chocolate Pudding by thermo-processing in a retort pouch.

SS CRANAPPLE DESSERT

2/3 cup canned pineapple tidbits in unsweetened pineapple juice
1/2 lb Golden Delicious apples, peeled and sliced
1/2 lb whole cranberries
1/2 cup rolled oats
1/4 cup blanched sliced almonds

1 tbsp modified food starch (cornstarch may be substituted)
1/4 tsp ground cinnamon
1/2 cup apple juice
1/2 cup light brown sugar
2 tbsp maple syrup
1 tbsp unsalted butter
1 tsp pure vanilla extract

1. Pour off juice from canned pineapple and discard.
2. Combine pineapple, apples, and cranberries. Mix well. Set aside.
3. Combine oats and almonds. Mix well. Set aside.
4. Combine starch, cinnamon, and 1 tbsp apple juice and mix well. Set aside.
5. Add the remaining apple juice, brown sugar, maple syrup, and butter to a saucepan and begin heating on medium heat.
6. Add the apple juice starch mixture while stirring.
7. Heat mixture to simmer and add vanilla
8. Add pineapple, apples, cranberries, oats, and almonds and heat to boiling. Simmer 5–7 minutes until apples and cranberries are tender.

Yield: Serves 6

Note: NASA further processes the Cranapple Dessert by thermo-processing in a retort pouch.

EMERIL'S MIXED FRUIT PANDOWDY

2 Golden Delicious or Granny Smith apples, peeled and sliced into 1/2-in. thick pieces

2 pears, such as Anjou, peeled and sliced into 1/2-in. thick pieces

4 nectarines or peaches, pitted and sliced into 1/2-in. thick pieces

2 plums, pitted and sliced into 1/2-in. thick pieces

1 cup strawberries, rinsed and patted dry, hulled, and quartered

1/2 cup blackberries, rinsed and patted dry

2 tsp fresh lemon juice

2 tsp cornstarch

1/2 cup plus 1 tbsp sugar or maple syrup

1/4 tsp ground cloves

1/8 tsp ground nutmeg

Pinch salt

1½ tbsp unsalted butter

1. Preheat the oven to 400°F.
2. In a large mixing bowl combine the apples, pears, nectarines, plums, strawberries, and blackberries with the lemon juice and cornstarch and toss to combine.
3. Add 1/2 cup sugar, cloves, nutmeg, and salt and stir well.
4. Butter a deep pie dish or a 9-by-12-in. baking dish with 1/2 tbsp of the butter. Transfer the fruit mixture to the buttered dish, and dot the top of the fruit mixture with the remaining tablespoon of butter.
5. Bake the fruit, uncovered, about 30 minutes. Remove from the oven. Let cool 15–20 minutes.

Yield: 6–8 servings

Note: When used in space NASA further processes the Mixed Fruit Pandowdy by freeze drying.

SS RICE PUDDING (FIGURE 8.4)

1 tbsp salted butter
2/3 cup golden raisins
1 tbsp rum extract
1/4 tsp fresh vanilla bean seed from
  the pod
1/2 cup uncooked long grain white
  rice

3/4 cup water
1/2 cup light brown sugar
1/3 cup egg substitute, such as Egg
  Beaters™
1/4 tsp salt
1 1/3 cups whole milk

1. Prepare one baking pan (11″ × 8½″ × 3″) by coating with butter. Preheat oven to 350°F.
2. Measure raisins, rum extract, and vanilla bean seed. Mix well and set aside.
3. Measure rice and water, and prepare per package directions. Cook rice approximately 20–25 minutes. Set aside to cool.
4. Measure brown sugar, egg product, and salt. Combine ingredients and mix well with a wire whisk.
5. Measure and add milk to this mixture.
6. Combine the milk mixture, raisins, and cooked rice in a container.

FIGURE 8.4 Rice pudding.

7. Pour the mixture into the buttered pan, using a large spatula to spread it evenly throughout the pan.
8. Bake for approximately 30 minutes. Remove the pan from the oven and serve.

Yield: 6 servings

Note: NASA further processes the Rice Pudding by freeze drying.

EMERIL'S RICE PUDDING WITH RUM RAISINS

3/4 cup golden raisins
2 tbsp rum extract
1 vanilla bean, split lengthwise
1 cup water
1/2 cup uncooked long-grain white
   rice

1 tbsp unsalted butter
1½ cups whole milk
1/2 cup light brown sugar
2 large eggs
1 large egg yolk
1/8 tsp salt

1. Place the raisins in a small bowl.
2. In a small saucepan heat the rum over medium heat. Pour the warm rum over the raisins, cover, and let soak at least 30 minutes and up to 2 hours. Drain.
3. Scrape the vanilla beans into a small ramekin and reserve.
4. Combine water, rice, and vanilla bean pod in a heavy, medium saucepan. Bring to a boil, then reduce heat to medium-low and simmer, covered, until the rice is tender and the liquid is absorbed, about 20 minutes.
5. Uncover rice, discard the vanilla bean, and let cool.
6. Preheat oven to 350°F.
7. Butter one large (6- to 8-cup) soufflé dish with the tablespoon of butter, and place inside a roasting pan.
8. In a large bowl whisk together the milk, brown sugar, eggs, egg yolk, salt, and reserved vanilla seeds. Stir in the raisins and 1½ cups cooked rice. Pour into the buttered dish.
9. Add enough hot water to the roasting pan to come halfway up the sides of the dish. Bake until the pudding is set in center and brown around edges, about 1 hour 5 minutes. Remove

the dish from the water and cool at least 15 minutes before serving.

Yield: 6 servings

Note: When used in space NASA further processes the Rice Pudding with Rum Raisins by freeze drying.

SS TAPIOCA PUDDING IN POUCHES

2 cups whole milk
1 cup water
2 tbsp quick cooking tapioca
3/4 cup extra fine sugar
1/4 tsp salt

2 tbsp modified food starch (you can substitute corn starch)
1¼ tsp liquid tapioca pudding flavor (This is a food service product; no readily available substitute.)

1. Add 1½ cups milk, water, tapioca, sugar, and salt to a saucepan and allow mixture to sit for 5 minutes.
2. Heat mixture to boiling and remove from heat.
3. Add remaining milk to the starch mixture, mix well, and add to the heated milk mixture.
4. Add tapioca flavor. Mix well, transfer to serving dishes, and chill.

Yield: 6 servings

Note: NASA further processes the Tapioca Pudding by thermo-processing in a retort pouch.

SS VANILLA PUDDING IN POUCHES

2 cups whole milk
1 cup water
3/4 cup extra fine granulated sugar
1/4 tsp salt

4 tbsp modified food starch
(substitute corn starch)
1¼ tsp vanilla

1. Add 1½ cups milk, water, sugar, and salt to a saucepan, heat mixture to boiling, and remove from heat.
2. Add remaining milk to the starch mixture, mix well, and add to the heated milk mixture. Add vanilla flavor.
3. Mix well, transfer to serving dishes, and chill.

Yield: 6 servings

Note: NASA further processes the Vanilla Pudding by thermo-processing in a retort pouch.

Following are some the astronauts' favorite dessert recipes.

KAREN ROSS'S* BANANA PUDDING

*Karen is the wife of astronaut Jerry Ross (see below). Karen works in the Shuttle Food Processing Facility, where the shuttle food is produced. She also assists with food preparation during the pre-flight quarantine period, and this recipe is a favorite dessert during that period of isolation.

1 cup sugar
2/3 cup all-purpose flour
dash of salt
4 cups milk
8 egg yolks, well beaten (reserve
   egg whites for meringue)
1¼ tsp vanilla extract

1 box Nilla™ wafers
5 or 6 (or more) bananas, sliced
8 egg whites at room temperature
dash of salt
1/4 tsp cream of tartar
1½ tsp vanilla extract
5 tbsp sugar

1. Combine 1 cup sugar, flour, and salt in top of double boiler or in a large saucepan.
2. In mixing bowl, add 1 cup milk to egg yolks; beat thoroughly to combine. Add remaining 3 cups milk, and beat well.
3. Gradually stir milk and egg yolk mixture into dry ingredients; blend well with each addition, crushing all lumps.
4. Cook over boiling water or medium heat, uncovered, stirring constantly until thickened.
5. Reduce heat and cook, stirring occasionally, for 5 minutes. Remove from heat, and add 1 ¼ tsp vanilla.
6. Spread small amount of custard over bottom of lightly greased 9″ × 13″ baking pan.
7. Arrange Nilla™ wafers to cover bottom of pan. Top with layer of sliced bananas. Pour about 1/3 of custard over bananas. Continue to layer wafers, bananas, and custard to make three layers of each, ending with custard.
8. Preheat oven to 350°F. Beat egg whites, dash of salt, cream of tartar, and 1 ½ tsp vanilla until stiff but not dry. Gradually add 5 tbsp of sugar, 1 tbsp at a time, and beat until stiff peaks form.
9. Spoon on top of custard, spreading to cover entire surface, and sealing well to edges. Bake at 350°F for 5–10 minutes, or until slightly browned.

Yield: 15 servings

Meet the Astronaut: **Jerry Ross, STS-61B, STS-27, STS-37, STS-55, STS-74, STS-88, and STS-110**

Colonel Ross received a Bachelor of Science and Master of Science degrees in Mechanical Engineering from Purdue University in 1970 and 1972. Colonel Ross graduated from the USAF Test Pilot School's Flight Test Engineer Course in 1976 and was subsequently assigned to the 6510th Test Wing at Edwards Air Force Base, California. He was selected as an astronaut in May 1980. Colonel Ross flew as a Mission Specialist on STS 61-B (1985), STS-27 (1988) and STS-37 (1991), was the Payload Commander on STS-55/Spacelab-D2 (1993), and again served as a Mission Specialist on the second space shuttle to rendezvous and dock with the Russian space station *Mir*, STS-74 (1995), the first ISS assembly mission, STS-88 (1998) and STS-110 (2002).

A veteran of seven spaceflights, Ross has over 1,393 hours in space, including 58 hours and 18 minutes on nine EVA's (spacewalks). He was the first human to ever be launched into space seven times.

PAULA HALL'S* SUPER LEMON BARS

*Former shuttle/ISS dietitian for 10 years who died of cancer in 2007.

Crust:

2 cups flour
1/2 cup powdered sugar

3/4 cup margarine

Filling:

4 eggs
1 tsp salt
1 3/4 cups sugar
1/4 cup lemon juice

1 tsp. baking powder
1 tsp lemon zest

Icing:

2½ cups powdered sugar
1 tbsp margarine

1/4 cup lemon juice
1 tsp lemon zest

For Crust:

1. Preheat oven to 350°F. Mix flour and powdered sugar.
2. Cut in margarine with a pastry blender until mixture is crumbly.
3. Press into 9″ × 13″ glass baking pan that has been sprayed with a non-stick cooking spray.
4. Bake for 15 minutes or until light brown.

For Filling:

1. Mix all ingredients for filling. Pour onto hot crust.
2. Return pan to oven and bake 15–20 minutes or until top of filling is lightly browned. Allow to cool to room temperature.

For Icing:

Mix all ingredients well and pour on cooled crust. Tilt pan until all of filling is covered with icing. Allow a minimum of one hour for the icing to set up.

Yield: 18 bars

MIKE MULANE'S (ACTUALLY, HIS MOTHER'S) COCONUT CREAM PIE

2 eggs, separated
1/2 cup sugar
5 tbsp flour
1 cup evaporated milk
1 cup water
1 cup flaked coconut

Pinch of salt
1 tsp vanilla extract
4 tbsp butter or margarine
1 pre-baked 9-in. pie shell
1/4 cup sugar
1/4 tsp cream of tartar

1. Beat egg yolks.
2. Combine yolks, ½ cup sugar, flour, milk, water, coconut, and salt in a medium saucepan; blend thoroughly.
3. Cook over medium heat, stirring frequently, until thickened. Remove from heat; stir in vanilla and butter.
4. Spoon mixture into pastry shell.
5. Beat egg whites until foamy. Gradually add in ¼ cup sugar and cream of tartar; continue beating until stiff peaks form.
6. Spread meringue over pie, being careful to seal edges.
7. Bake at 400°F about 10 minutes, or until lightly browned. Cool. Refrigerate until serving time.

Meet the Astronaut: **Mike Mulane, STS-41D, STS-27, and STS-36**

Mike Mulane received a Bachelor of Science degree in Military Engineering from the United States Military Academy in 1967 and a Master of Science degree in Aeronautical Engineering from the Air Force Institute of Technology in 1975. He was selected by NASA in January 1978 and became an astronaut in August 1979. A veteran of three spaceflights, he has logged a total of 356 hours in space. He was a Mission Specialist on the crew of STS-41D (August 30 to September 5, 1984), STS-27 (December 2-6, 1988), and STS-36 in (February 28 to March 4, 1990). Mike is currently a professional speaker and writer and is the author of a book entitled "Riding Rockets" (Scribner 2006) in which he describes what it is like being an astronaut.

CONNIE STADLER'S* RHUBARB MUFFINS

*Former *Apollo*, *Skylab*, and shuttle dietitian from 1970 to 1988.

Muffin:

1/2 cup sugar
1 cup rhubarb, chopped
3/4 cup Bisquick™
1 egg

1/2 cup sour cream
1 tsp vanilla
Dash of salt

Topping:

¼ cup butter
½ cup flour

½ cup sugar
¼ cup pecans, chopped

1. Mix sugar, rhubarb, and Bisquick together.
2. Mix lightly and add egg, sour cream, vanilla, and a dash of salt.
3. Spoon batter into muffin tins lined with paper liners. Fill each half full.
4. Mix topping ingredients together and sprinkle over batter, pressing down slightly.
5. Bake at 375°F for about 25 minutes, but watch carefully.

Batter keeps well a couple of days in the refrigerator (topping, too)

MILDRED BONDAR'S (ASTRONAUT ROBERTA BONDAR'S MOTHER) DREAM CAKE (FIGURE 8.5)

Shortbread:

1/2 cup butter

1/4 cup brown sugar

1 cup flour

Pinch salt

Filling:

2 eggs

1 cup brown sugar

1 cup desiccated coconut

1 cup finely chopped pecans

1 tbsp flour

1 tsp vanilla

1/2 tsp baking powder

Pinch salt

FIGURE 8.5 Astronaut Roberta Bondar removing her Dream Cake from the food locker on STS-42. The plastic bag containing the Dream Cake can be seen in the Imax movie, Destiny in Space (NASA photograph).

Icing:

1½ cups sifted icing sugar　　　1 tsp vanilla
1 tbsp butter　　　　　　　　　Milk to moisten

1. Preheat oven to 325°F.
2. In a bowl mix by hand butter and brown sugar, then add flour and salt.
3. Place mixture into a lightly greased pan (approx 7″ × 7″ × 2″) and bake.
4. For the filling, in a bowl, beat the 2 eggs and brown sugar and stir in remaining filling ingredients.
5. Pour mixture over shortbread and return to the oven for 30–40 minutes.
6. Remove from oven and cool.
7. Blend icing ingredients and mix with enough milk to moisten and spread on cool cake.
8. When icing has set, cut into squares.

Meet the Astronaut: **Roberta Bondar, STS-42**

Roberta Bondar received a B.Sc. in zoology and agriculture, University of Guelph, 1968, M.Sc. University of Western Ontario, 1971, Ph.D. in neurobiology, University of Toronto, 1974, and a MD from McMaster University, 1977. Dr. Bondar was one of the six original Canadian astronauts selected in December, 1983 and began astronaut training in February, 1984. In early 1990, she was designated a prime Payload Specialist for the first International Microgravity Laboratory Mission (IML-1). Dr. Bondar flew on the space shuttle Discovery during Mission STS-42, January 22-30, 1992 where she performed life science and material science experiments in the Spacelab and on the middeck.

What You'll Find at Your Supermarket

Don't feel like measuring, blending, folding, cooking, and baking space deserts? When nobody's looking, head to your local supermarket. Except for the packaging, the following is a list of authentic space deserts to please any sugarholic.

Spiced apple pieces in a retort pouch by SOPACKO
Handi-snacks™ Banana Split Pudding by Kraft
Individual quick frozen Hill Country Fare berry medley by HEB.
   NASA adds sugar and freeze dries before use.
Blueberry raspberry yogurt in pouches by Arla Foods
Handi-snacks™ butterscotch pudding by Kraft
Fruit Cocktail in retort pouches by SOPACKO
Mocha Yogurt in pouches by Arla Foods
Peaches in retort pouches by SOPACKO
Pears in Extra Lite Syrup by Del Monte
Pears in retort pouches by SOPACKO
Pineapple tidbits in pineapple juice by Del Monte
Pineapple in retort pouches by SOPACKO
Strawberries halved in delicious syrup by Birds Eye. NASA adds
   ascorbic acid and freeze dries before using.
Handi-snacks™ vanilla or tapioca pudding by Kraft

CHAPTER 9

Beverages

Millions of adults grew up drinking orange flavored Tang^{TM} at breakfast. In a widespread advertising campaign, parents learned that NASA astronauts drink Tang^{TM} in space, and therefore, it's good for children. Tang^{TM} became famous, and many attributed its invention to the space program. It was one example of how exploring space benefited people on Earth. The only thing true about this story, though, is that astronauts did drink it in space.

During the *Gemini* program, onboard water was produced as a byproduct in fuel cells that combined hydrogen and oxygen in an electrolyte solution to generate electricity. Since water was being created and it was harmless to drink, why not save weight and use it instead of launching large water tanks? The only problem was that the water processed by fuel cells had an unpleasant taste. Something was needed to mask its flavor, and that something was Tang^{TM}. Tang^{TM}, invented in 1957 by the same person that invented Pop Rocks, was a convenient powder that could easily be rehydrated in space. Although not nearly as good tasting as real orange juice, it did solve the spacecraft water taste problem. The rest was history.

To prepare Tang^{TM} for spaceflight, it must be packaged under a hard vacuum. Any air in the package would form bubbles in the drink during rehydration. If air gets into a beverage package in space, it is impossible to get out. The astronaut ends up swallowing the air, which causes intestinal

C.T. Bourland, G.L. Vogt, *The Astronaut's Cookbook*, DOI 10.1007/978-1-4419-0624-3_9,
© Springer Science+Business Media, LLC 2010

problems. One of the interesting side effects of spaceflight is that it is difficult to burp in space. Often, a burp is accompanied by a liquid spray (very unpopular). Dr. Joseph Kerwin, from the *Skylab 2* mission, once spoke to a Congressional panel and described the medical state of astronauts. He explained that the GI tract "accelerates bubbles downward with great velocity."

Bubbles are an occasional problem in beverages. Sometimes bubbles enter the water supply from the fuel cells. A hydrogen gas removal system usually takes care of the problem.

Powdered drinks seem to work best in space. Natural fruit juices will solidify when subjected to a hard vacuum over long periods of time. Reconstituting them with water is next to impossible. The only natural juice currently flown in space is freeze-dried orange and grapefruit juice.

Astronauts give up many comforts of home when they embark on space travel. Besides the loss of privacy, the comfort of a familiar bed, and home cooked meals, they leave behind carbonated soft drinks! In microgravity, sodas do strange things. Think about pouring a glass of Coke on Earth. The Coke fills the glass and then a head forms from bubbles of carbon dioxide that come out of solution and rise to the top. If you could pour a Coke in space (liquids don't pour in microgravity), the bubbles would not rise to the top of the glass. They would just expand and remain where they are. In no time, a bubbly froth would appear that would rise out of the glass—a big-time mess! Soda and another popular astronaut drink, beer, are undrinkable in space.

Regardless of the carbonation problem, soft drinks have flown in space. During the STS-51F space shuttle mission in 1985, both Coca Cola and Pepsi Cola were flown. The Space Food Laboratory personnel called it the great "Space Cola War."

It all started out when Coca Cola engineers developed a Coke dispenser in the shape of a Coke can. The effort cost a couple of hundred thousand dollars. Coca Cola wanted to test the dispenser in space. After submitting it through a detailed flight approval process, NASA granted permission for it to fly it as an experiment. So far so good. Then, Pepsi got wind of what was happening and demanded equal time. Both Coke and Pepsi would fly together in a "cola challenge."

The two experiments moved into the political arena, and NASA began receiving conflicting directions from Congress and the White House. The manager of the Space Food Laboratory had not once, in his 15 years at NASA, been summoned to the JSC Center director's office, but he soon found himself wearing a path in the director's carpet to periodically receive new instructions. Remove the brand labels from the dispensers and place NASA insignias on them. Remove the NASA insignia and replace the brand labels. Repeat. And so on. The final directive was to go with the brand labels, but astronauts were warned about making any comments that would inflame the war. In the end, the astronauts reported that both Coke and Pepsi tasted just like warm soda (there wasn't any refrigeration on board)! (Figure 9.1)

Even if a refrigerator were on board and a microgravity soda dispenser were perfected, there is no guarantee that carbonated beverages would be used in space. There is still the problem of the effects of bubbles on astronaut digestive systems.

OK, no soda and no beer. What about wine? Alcohol has never been an official part of any NASA Space Food System, although it nearly became a part of the *Skylab* system. Since the *Skylab* space station was planned to be a home away from home, some felt it

FIGURE 9.1 Astronaut Anthony England testing the experimental Coke can on STS-51F (NASA photograph).

should be stocked like a home. That included wine. After consulting with several professors at the University of California at Davis, it was decided that a Sherry would work best because any wine flown would have to be repackaged. Sherry is a very stable product, having been heated during the processing. Thus, it would be the least likely to undergo changes if it were to be repackaged.

The winner of the space Sherry taste test was Paul Masson, California Rare Cream Sherry. A quantity of this Rare Cream Sherry was ordered for the entire *Skylab* mission and was delivered to the Johnson Space Center. A package was developed that consisted of a flexible plastic pouch with a built-in drinking tube, which could be cut off. The astronaut would simply squeeze the bag and drink the wine from the package. The flexible container was designed to be fitted into the *Skylab* pudding can.

In the end, the flight set of wine never reached the packaging stage. NASA was afraid of negative publicity. Furthermore, a survey of the *Skylab* crews indicated that wine was not a high priority for them. Wine was permitted for the *Skylab* pre-flight and post-flight food programs, and the dedicated prime and backup crews made sure none of the Paul Masson Rare Cream Sherry was wasted.

On the ISS there is a limited amount of alcohol that is brought up by the Russians. The official Russian position is that they do not allow alcohol in space, but some of the cosmonauts have friends at the launch center.

How do you consume beverages in space? Beverages have to be in enclosed containers. A cup does not control liquids, as it does on Earth. Raising the cup to the mouth gets the liquid moving, and it all leaves the cup when the cup reaches the lips. An elongated bag holds dry ingredients, such as Tang^{TM}, or one of many powdered drinks including coffee, tea, and cocoa. Hot or cold water is injected into the package, and the beverage is consumed through a straw.

CHARLES BOURLAND'S DIARY: WINE IN SPACE?

My boss was Mormon and consequently, the job of heading the wine selection process for the *Skylab* missions fell to me. Selecting a wine was an interesting project for the people in the food laboratory, and we had no shortage of volunteers for the taste panel. To make sure wine was safe to fly in space, we even took some up in NASA'S KC-135 low gravity plane. The plane, a retired air force tanker, flies over the Gulf of Mexico and makes about 40 parabolic arcs that simulate the microgravity of space. For approximately 25 seconds, the passengers inside feel weightless. Because of its gut-wrenching effects, the plane has earned the nickname of the 'Vomit Comet.' (Figure 9.2)

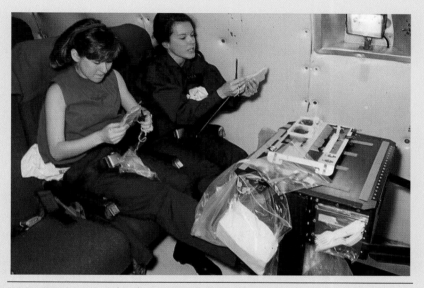

FIGURE 9.2 Food Lab personnel, Jane McAvin and Gloria Mongan test food packages on the zero G plane (NASA photograph).

The idea of flying on the KC-135 was to test the wine packaging in microgravity. As it turned out, the odors released by the wine, combined with a residual smell of years-worth of people getting sick on the plane, had an unplanned effect on the crew. Many grabbed for their barf bags. Usually, you have to drink a lot more wine before you get sick. Not here! In the end, the wine never flew on *Skylab*.

SPACE FIREWORKS

Due to power needs on board the space shuttle, hard-working fuel cells produce almost 2 gal/h of drinkable water as a byproduct. That's more than the astronauts can consume and more than cooling systems need. Eventually, when storage tanks fill up, excess water has to be gotten rid of. The excess is sprayed into space, whereupon it freezes into ice crystals. Astronauts report the crystal shower is like a fireworks show if the lighting is right. Sometimes, water dumps can even be seen from the ground, forming a small comet-like tail coming from the shuttle. When the shuttle is docked to the ISS, excess water is transferred to it.

CHARLES BOURLAND'S DIARY: WHAT YOU SEE IS WHAT YOU DRINK

Astronauts objected when the space shuttle's beverage package was changed from a plastic box with a clear plastic lid to a flexible foil container. The boxes didn't compress well in the onboard trash containers, and the new package would save lots of space. The astronauts objected because the contents were not visible. The objections died down when I pointed out that they couldn't see what was in a Coors or Coke can and that didn't stop them from drinking those products.

DRINKING FROM THE CUP

NASA was still using the space shuttle rigid square packages with flexible tops for beverages when astronaut Bill Thornton conducted a little impromptu flight experiment. He wanted to try to drink from a cup in space. Since there weren't any cups on board, Thornton filled a drink box with water and then cut off the top and had a drink. He raised the box to his mouth and started drinking, but then the rest of the water kept coming. Moving the box got all of the water moving as well, and it left the cup and ended up all over his face.

SUIT(ABLE) DRINKS

Today's spacesuits have a mechanism for allowing an astronaut to drink fluids while wearing the suit. A pouch inside the suit contains the fluid, and the astronaut drinks through a tube with a bite valve on the end. He or she bites the valve to open it and then sucks the fluid into the mouth. On *Apollo 16* it was decided to include an orange drink spiked with potassium gluconate because the astronauts on the previous Moon mission had heart arrhythmias, and the medical folks had decided that these were due to decreased potassium levels. It was felt that consuming potassium while walking on the Moon would eliminate the problem. While *Apollo 16* astronaut Charlie Duke was walking on the Moon, the bite valve became entangled with some communication wire and opened, releasing the potassium-spiked orange drink inside his helmet. Although no harm was done, this could have been catastrophic. Needless to say, since then nothing but water has ever been placed inside the suit for drinking.

SS CHOCOLATE INSTANT BREAKFAST

3 tbsp Chocolate Ovaltine™ (in powder form)
1/4 cup non-fat dry milk

1½ tsp unsweetened cocoa
1 cup hot or cold water

1. Add dry ingredients to a resealable plastic bag.
2. Add water (hot or cold) and mix well.
3. Drink with a straw.

Yield: 1 serving

Note: NASA packages the dry ingredients in the beverage package and astronauts add water before consuming.

SS ORANGE GRAPEFRUIT DRINK

2½ tsp Orange drink powder, such   1 cup cold water
    as Tang™
2 tsp Grapefruit drink powder juice,
    such as Tang™

1. Add dry ingredients to a resealable plastic bag.
2. Add water and mix well.
3. Drink with a straw.

Yield: 1 serving

NASA packages the dry ingredients in the beverage package and astronauts add water before consuming.

SS ORANGE MANGO DRINK

3 tsp Orange drink powder, such as   1 cup cold water
    Tang™
2 tsp Mango drink powder, such as
    Tang™

1. Add dry ingredients to a resealable plastic bag.
2. Add water and mix well.
3. Drink with a straw.

Yield: 1 serving

Note: NASA packages the dry ingredients in the beverage package and astronauts add water before consumption.

SS MANGO-PEACH SMOOTHIE

2/3 cup Safari Mango Smoothie Mix (from America's Classic Foods)
1/2 cup Smoothie'O non-dairy instant frozen yogurt mix (from Gold medal Products Co.)

2½ tbsp Sahara Burst peach Dry Crystals drink mix (from Sysco)
4½ cups chilled water

Add mango smoothie mix, instant frozen yogurt, and peach drink mix to the chilled water in a blender and blend. Astronauts do not have a blender in space, so if you want to mix it like astronauts do, add the ingredients to a plastic bag and shake well before serving.

Yield: 6 servings

For use in space these are packaged in the shuttle beverage package. The powder is added to the package before sealing the top with the drink adapter sealed inside under a hard vacuum. The drink adapter contains a septum sealed in with foil. When the astronaut rehydrates the package the galley needle punctures the foil seal and the septum and directs water to the package. When the needle is withdrawn, the septum reseals the package so there is no leakage. The astronaut inserts the plastic straw into the septum after rehydration and is then able to drink the beverage. The straw has a clamp to seal it off when not in use. Otherwise the liquid would siphon out of the straw in microgravity and end up in a big glob at the end of the straw (Figure 9.3)

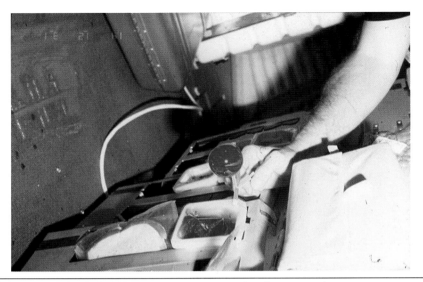

FIGURE 9.3 The STS-5 crew demonstrates what happens when straw is not clamped shut on the rigid beverage container (NASA photograph).

What You'll Find at Your Supermarket

Alpine Spiced Cider original apple flavor drink mix by Continental Mills, Seattle WA

Kool Aid™ brand Cherry Sugar free low calorie drink by Kraft

Cocoa MRE

Nescafe Taster's Choice™ Instant by Nestle. Add Coffeemate to Instant Nescafe to get coffee with cream

Grape Drink-Kool Aid™ brand grape artificial flavor sugar sweetened drink mix by Kraft

Grapefruit Drink- Crystals brand grapefruit freeze dried powder by Mastertaste, Plant City FL (add 1 teaspoon sugar)

Kool Aid™ brand grape sugar free low calorie drink by Kraft

Gatorade™ Lemon Lime drink powder

Country Time lemonade flavor drink by Kraft

Wyler's Light Lemonade Soft Drink Mix by Jel Sert

Lemon-Lime AstroAde by Shaklee Corp

Orange AstroAde by Shaklee Corp

Tang™ orange drink mix by Kraft

Crystals brand orange freeze dried fruit powder by Mastertaste

Tang™ pineapple flavor drink mix

Peach Apricot Drink-Sahara Burst peach dry crystals mix by Sysco
Tang^{TM} pineapple flavored drink mix by Kraft
Crystal Light^{TM} raspberry-peach mix by Kraft
HE Buddy strawberry instant drink mix by HEB
Tang^{TM} strawberry flavor drink mix by Kraft
Lipton diet iced tea mix with natural lemon flavor by Lipton
Kool Aid^{TM} tropical punch sugar sweetened soft drink mix by Kraft
Kool Aid^{TM} sugar free tropical punch mix by Kraft
Kroger brand Vanilla Instant Breakfast

CHAPTER 10

Future Space Food

The time has come for humankind to return to the Moon and to start preparing to make the first manned voyages to Mars. NASA's 21st Century Space Policy calls for just those steps, but going to the Moon won't consist of *Apollo*-style flights (grab some rocks and go). Astronauts will go there to build a permanent base. Eventually, the same is planned to happen on Mars. Making this vision possible will be a new family of space vehicles that combine the best technology of the past with the best of the future. Looking like a Saturn V rocket on the launch pad, the Ares I rocket will launch the *Orion* capsule with as many as six astronauts to the ISS or four astronauts for flights to the Moon. The Ares V, a much beefier rocket, will carry heavy payloads into orbit. For Moon missions, Ares V will loft the *Altair* lunar lander and the Earth Departure Stage (EDS) into orbit. After rendezvousing with *Orion*, the EDS will propel Orion and the lander to the Moon.

Although the first missions to the Moon are projected to be short and will not require significant, if any, modifications to the food system, the initial trip to Mars using current propulsion technology is projected to be a 2-year roundtrip. Travel to Mars, and back to Earth, require the two planets to be in favorable positions in their orbits. This means a 6-month flight out to Mars, a year on the surface, and a 6-month flight back.

C.T. Bourland, G.L. Vogt, *The Astronaut's Cookbook*, DOI 10.1007/978-1-4419-0624-3_10,
© Springer Science+Business Media, LLC 2010

Therefore, a Mars mission will require a food system with far longer shelf life (2–3 years) than is currently available for the ISS missions.

The challenges for a future food system will be very similar to the challenges presented by all previous space missions. Mass and volume of the foods system and their associated packaging will be limited. Refrigerators and freezers may not be available. Acceptability of the food items will become even more crucial on a 2-year mission. (After all, when you are 250 million miles from Earth, you can't just send out for a pizza. If you don't have a pizza maker on board, it is best not to think about pizzas.)

One of the new challenges for interplanetary missions is that food will have to be resistant to degrading due to exposure to radiation. Missions to the Moon and to Mars will carry astronauts well beyond the protective cocoon of Earth's Van Allen radiation belts. Food for the trip will not only be called upon to be nutritious but also to provide extra antioxidants to help counter the effects of radiation on the crew itself.

The future lunar habitat will be used to test technologies needed for a mission to Mars. Some of the plans for habitats include the growing of plants. Plants offer definite advantages in a closed environment, because they convert carbon dioxide to oxygen, recycle water, make food, and bolster crew morale. Unfortunately, the mass and size of the equipment needed for space agriculture—growth chambers, water and air recycling systems, and food processing equipment—make them impractical for short missions.

Nevertheless, through many short-term flight experiments, much has been learned about growing plants in space. American astronauts and Russian cosmonauts have flown plant experiments in space on the space shuttle, the *Mir* space station, and the ISS. Plants grow well in space, but goofy things can happen. Without a dominant direction for gravity, as is the case in orbital spacecraft, roots sometimes grow upward out of the soil. Air has to be deliberately circulated around plants. Lighting and temperature has to be carefully monitored and water limited so that plants do not become waterlogged. These are all manageable challenges. Crew members have reported that tending plants in space was a pleasurable experience, in part because plants provided a pleasant break from what is an otherwise relatively sterile environment.

NASA-sponsored research projects at various colleges have established that bell peppers, cabbage, carrots, beans, lettuce, green onions, herbs, peanuts, potatoes, radishes, rice, soybeans, spinach, strawberries, sweet potatoes, tomatoes, and wheat are all potential candidates for successful space agriculture. Some of the research projects focused on developing food products such as bread, cookies, and beverages from sweet potatoes. Recipes and menus were developed using only these crops as the food source, along with a few selected spices.

Other plant projects focused on growing technologies or on breeding plants with space-friendly properties. Using nutrient recycling systems, controlled lighting, optimum temperatures, and an atmosphere with the right mix of carbon dioxide, a sustainable diet for a single astronaut could be produced using only about 30 m^2 of growing space.

Amazing results such as this depend in part upon special dwarf varieties of plants such as dwarf wheat, rice, and tomatoes. Bred specifically for space flight, dwarf plants are short (naturally), and most of their growth goes into the production of grains or fruit. Furthermore, the plants thrive on 24 hours of light and carbon dioxide-rich atmospheres.

Needless to say, a plant-based advanced life support system results in a vegetarian diet. If you volunteer for one of the future extended space missions, you should know that vegetarian meals are what you will get.

However, you will have one big advantage over growing plants in microgravity if you are on the Moon or Mars. On these bodies there is sufficient gravity for plants to grow normally. Crops can more easily be processed into ingredients for other foods (i.e., wheat milled into flour). Envisioned is some sort of food synthesizer that would raise plant and animal cells and then process them into basic or even gourmet food. A computer-controlled food production machine that could use a few basic raw ingredients to make truly tasty and varied foods has been envisioned in science fiction and may prove to be successful someday in the future. "Computer, a cheese and sausage pizza, extra cheese, no anchovies" may be heard in other parts of the Solar System as well as here on Earth.

One thing NASA is very good at is simulation. No astronaut rockets into space without rehearsing every moment of the mission from liftoff to landing, including food preparation and dining. Astronauts practice everything. A permanent base on the Moon and a mission to Mars will pose unique challenges for the crew. Simulations for these missions have already begun.

In the late 1990s, volunteers at the Johnson Space Center were sealed inside a multi-story metal chamber for periods of up to 91 days. The chamber looked like a giant pressure cooker. The volunteers lived in tight quarters in a good approximation of the interior of a lunar or Mars base. The chamber was next to but not directly connected by airlocks to chambers where plants were grown. Oxygen produced by plants in the chambers was cycled into the crew chamber. During some of their studies nearly all of their air and water was recycled. For most of the experiments, the volunteers ate an ISS-style diet. In one experiment, they followed a limited vegetarian diet for 10 days that was definitely not "all you can eat." Regardless, the volunteers thrived, and much valuable data and insights were gained. A typical daily menu for the ten-day vegetarian part of the habitation experiment included:

Breakfast	Lunch	Dinner
Plain bagel	Vegetable chowder	Spaghetti sauce
Strawberry jelly	Spicy black bean burger	Whole wheat spaghetti
Orange juice	Soy bread	Skillet garlic bread
Coffee	Apricots	Cooked spinach
	Peanut butter pie	Salad with tomatoes and onions
Snack (Sugared Beverage)	Beverage	Peanut butter pie
	Snack (Pretzel sticks)	

Thanks to NASA-sponsored university and college researchers (professors and students) and the dietitians at the Food Laboratory at the Johnson Space Center, you, too, can imagine yourself as a future Moon or Mars explorer sitting down to a well-deserved meal after a long day in a spacesuit exploring an alien world. Following is a sampling of the space food vegetarian recipes that have been developed for future astronauts.

CHARLES BOURLAND'S DIARY: BACK HOME SAFE

I was on the recovery ship for the famous *Apollo 13* mission that had to be aborted in mid-flight due to an explosion. The crew that flew used the Moon's gravity to turn their capsule back to Earth and barely made it home.

Once the capsule was on deck, I was able to poke my head into the hatch and look around. To my surprise, the capsule reeked with the odor of 8-hydroxyquinoline sulfate. I was expecting lots of other odors. The compound was contained in a pill that was placed in emptied food packages to prevent microbial growth on any leftover food residue (Figure 10.1). Although the smell was irritating to me, the crew never complained about it, probably becoming accustomed to it over time. Later research determined that the compound was not needed, and it was eliminated from future food packages beginning with the *Apollo/Soyuz* mission in 1975.

FIGURE 10.1 Apollo pill used to prevent spoilage of food residue (NASA photograph).

LUNAR TRIVIA

What was the first meal eaten on the surface of the Moon? *Apollo 11* astronauts spent 24 hours on the Moon before rocketing back into orbit to join crewmate Michael Collins. After a dramatic touchdown, in which Armstrong had to propel the lander safely across a boulder-strewn plain to find a clear landing spot and before making their first steps on the surface, the two lunar explorers had a celebration meal. So what was their first meal on the Moon? Bacon squares, peaches, sugar cookie cubes, pineapple grapefruit drink, and coffee.

FOOD OF THE FUTURE?

In the early 1980s a youth organization called the Young Astronauts was formed to inspire youngsters to excel in science, mathematics, and technology. At a national convention of Young Astronauts' chapters, the headquarters staff unveiled a new space-type nutritional food bar they called "food of the future." The idea was that chapters would offer the bars as a snack during and after school meetings. The YA Council would earn some money from the sales of the bars, and the kids would get a fun nutritional snack. To launch the product, a table was heaped with free bars for the Young Astronaut attendees to sample. Through the clear wrap, the bars closely resembled the seed bars hung in birdcages. Almost to a child, the free bars were rejected untouched—a valuable lesson in space food development. Even for gung-ho future space explorers, food presentation was still important.

BBQ TEMPEH*

8 oz tempeh* (available in larger
    supermarkets)
3 tbsp vegetable oil
1/4–1/2 cup water
1 tsp salt
1 cup chopped green onions
2 cloves garlic, minced
1/4 cup vegetable oil
2½ cups canned tomato sauce
1/4 cup water

1/2 cup brown sugar
1 tbsp molasses
1/2 cup mustard
1 tsp salt
1 tsp allspice
2 tsp crushed red pepper
2 tbsp minced parsley
1/4 cup water
1 tbsp soy sauce
1/4 cup lemon juice

*Tempeh is a high protein food made from cooked and slightly fermented soybeans and formed into a patty, similar to a very firm veggie burger. Many commercially prepared brands today add other grains, such as barley or rice, or spices to vary the flavor and nutritional content.

1. In a skillet fry squares of tempeh in oil on one side until golden brown.
2. Add water to cover and salt and put a lid on immediately.
3. When the water steams away, flip the squares, add more water, and fry and steam on the other side.
4. In a separate pan, sauté chopped onion and garlic in oil until onions are soft.
5. Add all ingredients except lemon juice and soy sauce.
6. Bring to boil, reduce heat, and simmer for 1 hour. Stir occasionally.
7. Add lemon juice and soy sauce and cook 10 minutes longer.

Yield: 4 servings

BRAISED ONIONS AND CARROTS (FIGURE 10.2)

4½ tbsp vegetable oil
1 medium onion, thinly sliced
1½ large tomatoes, peeled and
    chopped
24 baby carrots, sliced lengthwise
1 tsp salt

1/4 tsp cayenne pepper
1/2 cup finely chopped scallions
3 tbsp finely chopped cilantro

1. Heat oil in a heavy skillet over high heat.
2. Drop in onions and, stirring frequently, cook over moderate heat for 8–10 minutes, or until golden brown.
3. Add tomatoes, increase heat, and boil briskly, uncovered, until most of the liquid has evaporated.
4. Stir in carrots, cilantro, salt, and cayenne pepper.
5. Add enough water (½–³/₄ cup) to barely cover the carrots.
6. Bring to boil, cover with lid, and simmer for 10 minute, or until carrots are tender.
7. Sprinkle with chopped scallions and serve.

Yield: 6 servings

FIGURE 10.2 Braised onions and carrots.

TOFU BROWNIES

1/3 cup unbleached all-purpose
   flour
2/3 cup water
1/2 lb tofu
2 cups sugar
1 tsp salt

1 tsp vanilla
3/4 cup cocoa
1/2 cup vegetable oil
1½ cup unbleached all-purpose
   flour
1 scant tsp baking powder

1. Preheat oven to 350°F.
2. Whisk the 1/3 cup flour and water in a cool saucepan until lump free.
3. Whip tofu in a blender until smooth and creamy.
4. Add flour and water to tofu and cook over low heat until mixture thickens. Remove from heat and allow to cool.
5. To cooled mixture add sugar, salt, and vanilla. Beat well to remove lumps.
6. In a second bowl mix cocoa and oil until lumps are gone, and add to the tofu/flour mixture.
7. Mix flour and baking powder and add to the mixture. Stir until lumps are gone.
8. Bake in a well oiled and floured 9″ × 13″ pan for 25 minutes at 350°F.

Yield: 16 bars

SOY BREAD (FIGURE 10.3)

3/4 cup +2 tbsp plain soymilk
1 tbsp peanut oil
2 tsp dry active yeast
2 tbsp sugar

1 tsp salt
1/4 cup + 2 tbsp soy flour
2 cups unbleached all-purpose flour

This recipe is for a 2-lb loaf to be baked in a bread machine. Follow the baking instructions for the bread machine.

Yield: 12 slices

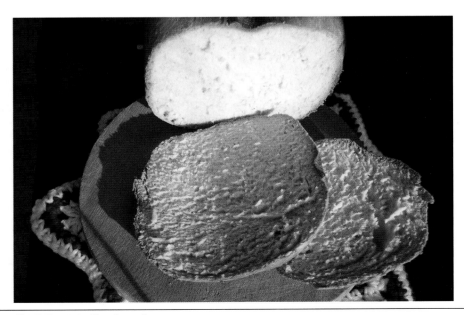

FIGURE 10.3 Soy bread.

CACCIATORE (TEMPEH)

16 oz tempeh
1/2 cup unbleached all-purpose
   flour
1 tsp salt

1/4 tsp pepper
1/4 cup water
3 tbsp peanut oil

Sauce:

1 29-oz can of tomato sauce
1 bay leaf
2 cloves garlic, minced
1/8 tsp thyme
1 tsp basil

1 tsp oregano
1/8 tsp marjoram
1 tbsp sugar
1 cup chopped green onion

1. Measure out all ingredients.
2. Cut tempeh into 12–18 short pieces.
3. Mix flour, salt, and pepper in a bag and set aside.
4. Steam tempeh by heating water in a skillet over medium heat. When water is hot, add tempeh and steam for 5 minutes, or until water evaporates.
5. Remove tempeh and place on dish; allow to cool for 5 minutes.
6. In a separate large soup-style pan, combine all sauce ingredients and heat to boiling. Allow to simmer on low heat.
7. Coat tempeh pieces in the bag of flour, salt, and pepper.
8. Heat oil on medium heat in skillet. When a small piece of tempeh will sizzle on being placed in the oil, add remaining tempeh and onions and brown until medium to dark brown.
9. Remove browned tempeh and pat excess oil.
10. Add browned tempeh and onions to the sauce and simmer on low to medium heat for 20–30 minutes.

Yield: 4–6 servings

ENGLISH MUFFINS

1 cup hot water
1/2 cup scalded soymilk
2 tsp sugar
1 tsp salt
1 package dry active yeast
2 tbsp warm water

4 cups unbleached all-purpose flour
3 tbsp butter flavoring such as Butter Buds™
3 tbsp vegetable oil

1. Combine water, milk, sugar, and salt in a mixing bowl.
2. Dissolve the yeast in the 2 tbsp warm water for 10 minutes.
3. Beat half of the flour into the milk mixture.
4. Cover the bowl with a damp cloth and put in a warm place for about 90 minutes, until dough rises.
5. Beat in butter flavoring, then beat in the rest of the flour and let the dough rise again for another 90 minutes.
6. Place the dough on a lightly floured board. Press or pat until the dough has doubled in area. Cut into circles with a cookie cutter or cup.
7. Cook dough circles until light brown on a fairly hot, well-oiled griddle. Turn each muffin once while cooking.

GARBANZO BEAN BURGERS

2 cups cooked garbanzo beans, mashed
1 stalk celery, finely chopped
1 carrot, finely chopped

1/4 cup minced green onion
1/4 cup whole wheat flour
2 tsp vegetable oil

1. Mix all ingredients except oil in a bowl. Form 6 flat patties.
2. Fry in oiled pan over medium-high heat until burgers are golden brown on each side.

Yield: 6 servings

HOT AND SOUR SOUP

4 cups water
1/4 lb tofu, firm, slivered
2 cubes vegetable bouillon
3/4 tsp salt
1 tbsp soy sauce

2 tbsp white vinegar
1/4 tsp cayenne pepper
3 tbsp water
2 tbsp cornstarch
1/4 cup green onions, chopped

1. Bring water to boil.
2. Add tofu, bouillon, salt, and soy sauce. Simmer for 3 minutes.
3. Add vinegar and cayenne pepper and bring to boil.
4. In a separate dish dissolve together 3 tbsp water and cornstarch. Add dissolved mixture to the soup. Garnish with chopped green onions.

Yield: 4 servings

KIDNEY BEAN BURGERS (FIGURE 10.4)

1 29-oz can of kidney beans,
   drained
2 cups cooked brown rice
1/4 cup chopped green onions
2 tbsp ketchup
1/2 tsp garlic powder

1 tsp oregano
1/8 tsp thyme
1/4 tsp sage
1/2 tsp salt
1/2 tsp pepper

Combine all ingredients in a large bowl and mash well. Form into patties and cook in a non-stick pan until browned on both sides.

Yield: 6 servings

FIGURE 10.4 Kidney bean burgers.

POTATO SALAD

2 cups diced cooked potatoes
1/2 cup diced celery
1 tbsp chopped green onions
1 red bell pepper, finely chopped
2 tbsp chopped fresh flat-leaf
  parsley

1/2 tbsp cider vinegar
1 tsp dry mustard
1/2 tsp celery seed
1/4 cup tofu-based mayonnaise
  (see recipe below)

Combine all ingredients except mayonnaise. Toss lightly and chill. A few hours before serving time, add mayonnaise and return to the refrigerator.

Yield: 4 servings

Tofu Mayonnaise:

½ lb tofu
¼ cup canola oil
1 tbsp lemon juice
1 tbsp sugar

1½ tsp bottled mustard
1 tsp apple cider vinegar
½ tsp salt

Combine all ingredients in a blender and beat until smooth and creamy.

PEANUT BUTTER BREAD

1 cup plain soymilk
2 cups unbleached all-purpose
  flour
1/3 cup sugar
2 tsp baking powder

1/4 tsp salt
3/4 cup creamy-style peanut
  butter, softened by stirring
1/3 cup liquid egg product such as
  Egg Beaters

This recipe is for a 1-lb loaf to be baked in a bread machine. Follow the instructions for the bread machine. Use the sweet bread mode. Add 1/3–½ dry ingredients to wet mixture in the loaf pan. This prevents the liquid from splashing during kneading. When the continuous kneading cycle (fast cycle) begins, add remainder of dry ingredients.

Yield: 12 slices

PEANUT BUTTER COOKIES (FIGURE 10.5)

3 cups unbleached all-purpose
   flour
1 tsp baking soda
1/2 tsp salt
1 cup brown sugar

1/2 cup vegetable oil
1 cup peanut butter
1/2 cup honey
1/2 cup tofu, blended
1 tsp vanilla

1. Preheat oven to 350°F.
2. Combine dry ingredients.
3. Combine wet ingredients.
4. Add flour mixture in 3 portions, stirring after each addition.
5. Form dough into 1-in. balls. Arrange on cookie sheet about 3 in. apart. Press with fork that has been dipped in cold water to create a crisscross design.
6. Bake for 10–12 minutes.

Yield: 48 cookies

FIGURE 10.5 Peanut butter cookies.

RICE-ORANGE PUDDING

3/4 cup rice, uncooked
1/4 tsp salt
1/2 cup sugar
1½ cups water
1½ tsp ground cloves

1 tsp grated orange peel
1/2 cup sugar
1 cup boiling water
1½ 11-oz cans mandarin oranges, drained

1. Combine rice, salt, 1/2 cup sugar, and water in a saucepan; cover, bring to boil, reduce heat, and simmer 15–20 minutes, or until water is absorbed and rice is tender.
2. Mix cloves, orange peel, sugar, and water in a small saucepan. Boil over medium heat for 5 minutes to form syrup.
3. Shape the rice into a mound on a platter. Arrange orange sections on rice and pour the syrup over them. Serve hot or cold.

Yield: 6 servings

ROASTED GARLIC SOYBEAN HUMMUS

3 large garlic cloves
2 cups cooked soybeans
2 tbsp sesame seeds, browned in oil
1 tbsp vegetable oil
1 tbsp vinegar
1/2 cup fresh parsley or cilantro, chopped

2 tsp salt
1 tbsp (1/2 cube) chicken bouillon
3 tbsp roasted bell peppers (canned or bottled are OK)
1–2 tsp cayenne pepper, to taste
1–2 tsp chili powder, to taste
1 tbsp cumin

1. Preheat broiler. Broil garlic cloves for 5–10 minutes, or until they just begin to brown and can be easily pierced with a fork.
2. In a food processor combine garlic with soybeans, sesame seeds, oil, vinegar, and parsley.
3. Blend until smooth and thick, adding salt, vinegar, and bouillon to reach desired consistency and flavor. Spread on crackers or veggies.

Yield: 1½ cups

SOYBEAN SOUP

1½ cups drained, pre-cooked
  soybeans
2 cups water
1¾ cup chicken broth
1/4 cup chopped carrot
1/4 cup chopped green onions
1/2 tomato, chopped

2 tsp dried thyme
2 tsp basil
1/2 tsp grated lemon zest
1 tsp lemon juice
Salt and pepper to taste
1/4 cup chopped celery
1 clove garlic, minced

1. Presoak soybeans.
2. Combine all ingredients and bring to a boil. Reduce heat and simmer 30 minutes, stirring occasionally, until vegetables are tender crisp.

Yield: 6 servings

SWEET AND SOUR TEMPEH

2–4 carrots, sliced
1 cup chopped green onions
1 bell pepper, cut into narrow strips
2 8-oz packages of tempeh
3–6 tbsp vegetable oil (for browning)
1/3 cup vegetable oil (for cooking carrots)

1/2 tsp garlic powder
1–2 tsp ginger powder
1 12-oz can of pineapple chunks (juice drained and reserved)
Cornstarch paste (2 tbsp cornstarch and 2 tsp water)

Sauce:

½ cup water
½ cup vinegar
2 tbsp soy sauce

6 tbsp brown sugar
Juice reserved from pineapple chunks

1. Chop all vegetables and tempeh before cooking.
2. Cut tempeh into 16–18 pieces per package.
3. Add water, vinegar, soy sauce, brown sugar, and juice and mix in a bowl.
4. Heat oil on medium in a skillet. As soon as a small piece of tempeh sizzles when placed in oil, add remaining tempeh.
5. Brown tempeh until medium to dark brown. Remove and pat dry to remove oil.
6. Heat 1/3 cup oil in a separate skillet on medium heat. Add garlic, ginger, carrots, and onions. Sauté until carrots are slightly soft. Add sauce ingredients and bring to a boil.
7. Mix cornstarch paste and add 1–3 tbsp sauce to the paste. Add this to the boiling vegetables and stir well. Reduce heat to low and add pineapple chunks and green peppers. Let cook until sauce thickens.
8. Serve tempeh on a bed of rice with sauce on top.

Yield: 4 servings

SWEET POTATO BREAD (FIGURE 10.6)

1 cup + 1 tbsp water
1 tbsp vegetable oil
2 tsp dry active yeast
¼ cup + 2 tbsp sweet potato flour*

3 cups unbleached all-purpose flour
4 tbsp sugar
1 tsp salt

*The sweet potato flour should be available on the Internet. If sweet potato flour is not available this dish can be made by dehydrating sweet potato slices in a food dehydrator and then blending the dried slices in a blender.
This recipe is for a 1½-lb loaf to be baked in a bread machine. Follow the instructions for the bread machine.

Yield: 12 slices

FIGURE 10.6 Sweet potato bread.

SWEET POTATO BISCUIT

1 medium sweet potato
6 tbsp vegetable oil
1/2 cup soy milk
1 tbsp sugar
1 egg beaten

2½ cups unbleached all-purpose
    flour
1 tbsp + 1 tsp baking powder
1 tsp salt

1. Preheat oven to 425°F.
2. In a medium saucepan, cook sweet potato in boiling water until tender when pierced with a knife, about 20 minutes.
3. Let the sweet potato cool, then peel it and mash until smooth.
4. In a medium bowl stir the mashed sweet potato with oil until smooth.
5. Stir in the remaining liquids and dry ingredients. Knead briefly in the bowl to form a soft dough.
6. On a floured work surface, roll the dough to $^3/_4$ in. thickness.
7. Using a cookie cutter, cut out biscuits. Transfer the biscuits to an ungreased baking sheet and bake 15–20 minutes.
8. Serve hot.

Yield: 12 biscuits

TOFU CHEESECAKE

2 lb tofu
1 cup sugar
1/4 cup lemon juice
1 tsp lemon extract

2 tsp vanilla extract
2 tbsp vegetable oil
1/4 tsp salt
1/2 cup honey

1. Preheat oven to 375°F.
2. Blend tofu in a blender until smooth and creamy.
3. Blend in sugar, lemon juice, lemon and vanilla extracts, oil, and salt. Blend in honey last.
4. Pour mixture into a crumb crust and bake for 40 minutes, or until small cracks start to form on the surface.
5. Serve well chilled.

Yield: 6 servings

WHOLE WHEAT BREAD

1 1/3 cup water
1 tbsp vegetable oil
1 tsp yeast
2 tsp salt

4 tbsp sugar
2 tbsp non-fat dry milk
3½ cups whole wheat flour

This recipe is for a 2-lb loaf to be baked in a bread machine.
Follow the instructions for the bread machine.

Yield: 12 slices

WHOLE WHEAT PANCAKES

1 cup whole wheat flour
1 tsp baking powder
1/4 tsp salt

1/4 tsp baking soda
1/2 cup soy or rice milk

Mix dry ingredients together; then mix in the soy milk. Cook in a
non-stick pan or on a griddle.

Yield: 9 pancakes

History of American Space Food

Mercury (1961–1963), *Gemini* (1965–1966) and *Apollo* (1968–1972)

Early US space food was highly engineered to minimize mass and volume and to prevent any possibility of food scraps contaminating the small cabins of the early NASA spacecraft. Space menu items consisted primarily of pureed foods in squeeze tubes, small cubed food items coated with an edible film to prevent crumbs, and freeze-dried powdered food items. It was agreed by most that early space food was, to say the least, unappetizing.

As the NASA manned space program progressed in the 1960's towards the first Moon landing in 1969, so did the food system. The *Gemini* program saw a food system that included formulation and packaging developments that were designed specifically for the program. Due to restrictions in weight and volume, concentrated foods were emphasized. Safety of the food system was highly emphasized, with the testing procedures developed for *Gemini* signaling the beginning of the Hazard

Analysis Critical Control Point (HACCP) program, which is now common practice for food safety around the world.

Apollo food systems introduced utensils to space dining with the addition of the 'spoonbowl' package, which allowed rehydrated food items to be consumed from the package with a spoon. The package had a wide zip-type opening and loops for a thumb and index finger. The loops permitted easy control of the package in microgravity.

During the *Apollo* program, the use of thermostabilized canned and pouched products in the US space program began. The real challenge for space food became apparent even this early in manned spaceflight—how to provide sufficient variety and quality to get the crew members to actually eat the food. Regardless of nutritional content, if the food was not consumed, the crew members' health was at risk. Crew members were returning from spaceflights with decreased body weight, and food was returned uneaten, indicating inadequate food consumption.

Skylab (1973–1974)

The *Skylab* space station of the mid-1970's actually featured the most sophisticated food system that NASA has ever flown in space to date. Frozen and refrigerated food items were included as part of the standard menu. To date, *Skylab* has been the only space project to have this luxury. The *Skylab* astronauts also had a dining table for meal consumption. Whether it was because of these advances or because of metabolic studies performed on *Skylab*, these crews had the highest percentage of actual versus planned food consumption for any US crews to date.

The Space Shuttle (1981 to the Present)

As a short duration work vehicle, the shuttle had inadequate space and not enough power to support refrigerator/freezers for food. NASA reverted back to an all shelf-stable food system. A meal tray was developed as a replacement for the bulky *Skylab* dining table. The original packaging for the shuttle food system used rigid plastic boxes with expandable lids for rehydratable foods and beverage items. As the program evolved and crew size and mission duration increased, the rigid packages were replaced with completely flexible packages that could be compressed to take up less space in the trash.

The shuttle food program adapted many commercially available food items for spaceflight. Although all items were repacked in spaceflight approved packaging, some were used as is (cookies, crackers, nuts), while others were further processed with freeze-drying (vegetables). The use of commercial items provided significant cost savings over the cost of developing unique foods for spaceflight. It also provided more familiar food items to the crew members. The use of commercial food items, however, has some disadvantages. Companies would sometimes change the contents ('New and Improved!') or discontinue items altogether. Furthermore, commercial products tend to be higher in fat, sodium, and sugar than preferred.

Besides expressing their preferences in choices offered, astronauts have contributed to the food system by recommending or actually bringing in favorite food samples for evaluation. Sally Ride, the first American woman to fly in space, contributed to the space food program by promoting personal preference menus. Hawaiian Ellison Onizuka, tragically lost in the *Challenger* accident, was responsible for adding Kona coffee and macadamia nuts to the shuttle food list. Mission Specialist Mary Cleave and Payload Specialist Rodolfo Vela introduced tortillas to the shuttle menu in 1985. Bruce McCandless, the first spacewalker to fly freely from a spacecraft in the Manned Maneuvering Unit, introduced trail mix. Including astronauts in the space food decision process (people who actually eat the food in space) has led to diversified menus and better-tasting food. Experience has shown that very few adjustments to the crew-selected menus are required to meet most nutritional requirements.

Shuttle/*Mir* Phase I Program (1995–1998)

Phase I of the International Space Station program saw US astronauts living for long periods of time on the Russian *Mir* space station with cosmonauts. Space shuttle and Russian space foods were consumed by astronauts and cosmonauts alike. As astronauts began to stay for long periods of time aboard *Mir*, the importance of food to crew members during the long months was magnified many times over that of short duration shuttle missions. It was apparent that food made significant psychological contributions to the crew. Post-flight

Mir crew debriefings revealed that thermostabilized items had far better long-term acceptability than their freeze-dried counterparts. It was also learned that increasing the variety of foods available to the crew members was important to prevent menu fatigue.

The International Space Station (November 2000 to the Present)

Using funds originally designated for the development of refrigerator/freezer-type foods (the planned refrigerator/freezer had been eliminated due to budget cuts), NASA food specialists began a new program to develop new shelf-stable food items. Product development focused mainly on thermostabilized food. The ISS does not create water with a fuel system, and so water is at a premium. Thermostabilized items, more acceptable to crew palate, also do not need water for reconstitution.

ISS food products were formulated to contain a lower level of sodium and fat over commercially available thermostabilized products. Although new product development focused on ISS needs, shuttle crews have benefited from the longer list of available foods for their missions.

The US food list currently consists of about 185 foods and beverages from which the shuttle and ISS crew members can build their menus. The Russian food list adds about 100 additional items to the total ISS food selection list.

Crews on the ISS eat a menu that is comprised equally of US space food and Russian space food. Meals are rotated. One day, meals consist of two American meals and one Russian meal plus a Russian snack. The next day, it is one American and two Russian meals and an American snack.

A 45-day contingency supply of food is maintained onboard ISS. If not used, it becomes part of the next crew's primary food, and new contingency foods are delivered by either a space shuttle or a Russian *Progress* vehicle. The contingency supply became especially important during the shuttle grounding following the loss of the space shuttle *Columbia* at the end of the STS-107 mission.

ISS flight menus are supplemented with a small quantity of 'bonus food' each month. The bonus foods are chosen by the crew and can be regular menu food items but often consist of commercially available candies, cookies, and crackers, not part of the standard ISS menu. On each shuttle and *Progress* flight, a small quantity of fresh food is stowed and transferred over to the ISS crew. This typically consists of fresh fruit (apples, oranges) and fresh vegetables (carrot sticks, onions, and garlic). Since refrigeration is not available, fresh items must be consumed quickly by the crew.

Because of the lengthy nature of the ISS missions, crew members settle into a more normal eating pattern than is possible on a hectic shuttle flight. For this reason, food consumption rates are much higher on the ISS but still not to the level observed during the *Skylab* program.

ISS Expedition Five (June to December 2005) Sample Crew Menu for Astronaut Peggy Whitson

Meals are designed on an 8-day rotating basis. Notice the alternating American and Russian menus. Russian mission planners insist on a 4-meal day. Meal 4 is actually a snack. The key to abbreviations is at the end of these charts.

DAY 1	DAY 2	DAY 3	DAY 4
Meal 1			
Chicken w/ Prunes (T)	Vegetable Quiche (R)	Omelet w/ Chicken (T)	Plain Yogurt (T)
Assorted Vegetables (R)	Beef Pattie (R)	Sweet Peas w/ Milk Sauce (R)	Seasoned Scrambled Eggs (R)
Cottage Cheese/Nuts (R)	Dried Peaches (NF)	Cottage Cheese/Nuts (R)	Cornflakes (R)
Wheat Bread Enriched (IM)	Shortbread Cookies (R)	Wheat Bread Enriched (IM)	Fruit Cocktail (T)
Kuraga (IM)	Kona Coffee w/ C&S (B)	Kuraga (IM)	Kona Coffee w/ C&S (B)
Coffee w/o Sugar (B)		Coffee w/o Sugar (B)	Pineapple Drink (B)
Vitamins		Vitamins	
Meal 2			
Jellied Pike Perch (T)	Mushroom Soup (R)	Appetizing Appetizer (T)	Split Pea Soup (T)
Borsch w/ Meat (R)	Tuna Noodle Casserole (T)	Peasant Soup (R)	Grilled Pork Chop (T)
Beef w/ Vegetables (R)	Pineapple (T)	Homestyle Beef (R)	Macaroni & Cheese (R)
Borodinskiy Bread (IM)	Candy Coated Peanuts (NF)	Moscow Rye Bread (IM)	Creamed Spinach (R)
Peach-Black Currant Juice (R)	Tea w/ Sugar (B)	Apple-Peach Juice (R)	Applesauce (T)
Tea w/ Sugar (B)		Tea w/o Sugar (B)	Tea (B)

DAY 1	DAY 2	DAY 3	DAY 4
Meal 3			
Lasagna w/ Meat (T)	Vegetable Spread (T)	Meatloaf (T)	Pike Perch in Baltica Sauce (T)
Pasta w/ Shrimp (R)	Meat w/ Barley Kasha (T)	Mashed Potatoes (R)	Chicken w/ Rice (T)
Asparagus (R)	Borodinskiy Bread (IM)	Italian Vegetables (R)	Borodinsky Bread (IM)
Dinner Roll (NF)	Russkoye Cookies (NF)	Shortbread Cookies (NF)	Visit Cracker (NF)
Lemonade (B)	Kuraga (IM)	Tea w/ L&S (B)	Prunes w/ Nuts (IM)
	Currant Tea w/o Sugar (B)		Green Tea w/o Sugar (B)
	Vitamins		Vitamins
Meal 4			
Dried Beef (IM)	Hazelnuts (NF)	Almonds (NF)	Sweet Almonds (NF)
Cashews (NF)	Apple Dessert (T)	Strawberries (R)	Kuraga (IM)
Peaches (T)	Peach-Black Currant Juice (R)	Lemonade w/ A/S (B)	Apple-Black Currant Juice (R)
Grape Drink (B)			

DAY 5	DAY 6	DAY 7	DAY 8
Meal 1			
Meat in Jelly (T)	Oatmeal w/ Raisins (R)	Chicken w/ Eggs (T)	Raspberry Yogurt (T)
Mashed Potatoes/ Onion (R)	Breakfast Sausage Links (I)	Buckwheat Gruel w/ Milk (R)	Scrambled Eggs (R)
Cottage Cheese/Apple (T)	Dried Apricots (IM)	Quince Bar (IM)	Oatmeal w/ Brown Sugar (R)
Wheat Bread Enriched (IM)	Strawberry Breakfast Drink (B)	Honey Cake (IM)	Kona Coffee w/ C&S (B)
Apple-Apricot Bar (IM)	Peach-Apricot Drink (B)	Coffee w/ Sugar (B)	Orange-Grapefruit Drink (B)
Coffee w/ Sugar (B)		Vitamins	
Vitamins			
Meal 2			
Spiced Pike Perch (T)	Split Pea Soup (T)	Bream in Tomato Sauce (T)	Beef Stew (T)
Sauerkraut Soup (R)	Grilled Chicken (T)	Pureed Vegetable Soup (R)	Rice w/ Butter (T)
Lamb w/ Vegetables (T)	Noodles & Chicken (R)	Veggie Ragout w/ Meat (R)	Wheat Flat Bread (NF)
Borodinskiy Bread (IM)	Broccoli au Gratin (R)	Moscow Rye Bread (IM)	Cauliflower w/ Cheese (R)
Apple-Black Currant Juice (R)	Shortbread Cookies (NF)	Peach-Apricot Juice (R)	Trail Mix (IM)

DAY 5	DAY 6	DAY 7	DAY 8
Tea w/o Sugar (B)	Orange-Pineapple Drink (B)	Tea w/o Sugar (R)	Apple Cider (B)
Meal 3			
BBQ Brisket (I)	Sturgeon (T)	Shrimp Cocktail (R)	Meat w/ Buckwheat Gruel (T)
Potatoes au Gratin (R)	Beef Goulash (T)	Chicken Fajitas (T)	Mashed Potatoes/ Onion (R)
Green Beans/ Mushrooms (R)	Stewed Cabbage (R)	Tortillas (NF)	Moscow Rye Bread (IM)
Pears (T)	Table Bread (IM)	Corn (R)	Milk (R)
Butter Cookies (NF)	Prunes w/ Nuts (IM)	Peach Ambrosia (R)	Apples w/ Nuts (IM)
Orange-Mango Drink (B)	Tea w/o Sugar (B)	Tea w/ Lemon (B)	Tea w/ Sugar (B)
	Vitamins		Vitamins
Meal 4			
Chicken Salad (R)	Hazelnuts (NF)	Cheddar Cheese Spread (T)	Visit Crackers (NF)
Crackers (NF) x2	Plum-Cherry Dessert (IM)	Crackers (NF) x2	Rossiyskiy Cheese (T)
Grapefruit Drink (B)	Grape-Plum Juice w/ Pulp (R)	Tropical Punch (B)	Apple-Plum Bar (IM)
			Apricot Juice (R)

B, beverage; IM, intermediate moisture; I, irradiated; R, rehydratable; T, thermostabilized; NF, natural form.

Internet Resources on Space Food and Nutrition

NASA Space Food Sites

http://spaceflight.nasa.gov/living/spacefood/index.html
http://spaceflight.nasa.gov/shuttle/reference/factsheets/food.html

NASA Educator Guide on Space Food and Nutrition

http://www.nasa.gov/audience/foreducators/topnav/materials/
  listbytype/Space_Food_and_Nutrition_Educator_Guide.html

National Public Radio Stories About Space Food

http://www.npr.org/templates/story/story.php?storyId=10792763

NASA Food Technology Commercial Space Center at Iowa State University College of Agriculture

http://www.ag.iastate.edu/centers/ftcsc/index.htm

Index

Printed in the United States of America